Reactor Operation

by

J. SHAW

Department of Nuclear Engineering,
Queen Mary College, London

THE QUEEN'S AWARD
TO INDUSTRY 1966

PERGAMON PRESS

OXFORD · LONDON · EDINBURGH · NEW YORK

TORONTO · SYDNEY · PARIS · BRAUNSCHWEIG

Pergamon Press Ltd., Headington Hill Hall, Oxford
4 & 5 Fitzroy Square, London W.1
Pergamon Press (Scotland) Ltd., 2 & 3 Teviot Place, Edinburgh 1
Pergamon Press Inc., Maxwell House, Fairview Park, Elmsford,
New York 10523
Pergamon of Canada Ltd., 207 Queen's Quay West, Toronto 1
Pergamon Press (Aust.) Pty. Ltd., 19a Boundary Street,
Rushcutters Bay, N.S.W. 2011, Australia
Pergamon Press S.A.R.L., 24 rue des Écoles, Paris 5^e
Vieweg & Sohn GmbH, Burgplatz 1, Braunschweig

Printed in Hungary

621.4 835
3534 r

Contents

Preface

MANY texts are available covering the general field of nuclear engineering, reactor design, reactor physics, reactor control, reactor instrumentation and associated topics. Reactor operation requires a wide knowledge of all these subjects with particular emphasis on the experimental and operational side. There is a demand for an introductory treatment of the problems associated with the commissioning and operation of nuclear reactors. It is the function of a reactor operations team to study the design specifications for a variety of items of plant and to bring the whole unit into operation. The team therefore acts in a coordinating capacity and merges the different items of equipment into an efficient operating system. Reactor operators must always be primarily concerned with the safety of the plant, the operations team and the general public. During operation all actions taken are based upon experimental results and previous experience associated with the known behaviour of a particular system. Design information is used only as a guide to reactor operation. Operations must not rely on theoretical information and due allowance must be made for errors in order to prevent a dangerous situation arising.

Thus the reactor operator tends to approach reactor problems in a way which is different to that of the reactor physicist or reactor designer. It is with this in mind that this text has been written. The field of reactor operation is very diverse and it is impossible in a text of this size to cover all problems encountered during operation, or indeed to follow any particular item in great detail. The aim has been to introduce the subject of reactor operation and to cover the main aspects.

For this reason all detailed descriptions have been restricted to the graphite-moderated gas-cooled reactors constructed during the initial phases of the United Kingdom nuclear power programme. It is felt that to avoid confusion it is better to consider one type of reactor in detail throughout the book when illustrating operational problems. It would be impracticable to deal in detail with more than one type of reactor. The treatment of the graphite-moderated gas-cooled reactor has been based on the author's own experience of the commissioning of the Berkeley reactors. When describing general operating principles an attempt has been made to cover other types of reactors within the scope of a text of this nature.

The book is intended for engineers and physicists working in the field of nuclear engineering and for undergraduate and postgraduate students requiring an introduction to the problems associated with the operation of reactors.

The book has been based on a course given to students studying for the M.Sc. degree in nuclear engineering at Queen Mary College (University of London).

CHAPTER 1

General Problems of Reactor Operation

1.1. INTRODUCTION

In 1939 at the Kaiser-Wilhelm Institute in Berlin, Hahn and Strassmann[1] studied the nature of the radioactive nuclei produced when uranium was bombarded with thermal neutrons. It was known at the time that a nuclear reaction resulted from the interaction of neutrons with uranium and that the product nuclei of the reaction were radioactive. It was thought that this induced activity was due to the formation of transuranium elements. It was something of a shock when the results of Hahn and Strassmann's investigations proved that a radioactive isotope of barium was produced as a result of the reaction. Lise Meitner and O. R. Frisch[2] suggested that a process took place, which they called fission, in which the uranium nucleus captures a neutron, becomes unstable, and splits into two halves of approximately equal weight. They also suggested that a large amount of energy would be released during such a process and this would appear as kinetic energy of the two fission fragments.

The discovery of the fission of the uranium nucleus was of tremendous importance as it opened up the prospects of using the energy stored in the atomic nucleus for the production of power from an entirely new and hitherto unknown source. At the time of the initial discoveries associated with the fission process, the future possibilities arising from this atomic

1

power seemed limitless. It can be calculated that the heat released as a result of the complete fission of 1 lb of uranium is equivalent to that obtained by the combustion of over 1200 tons of coal. A great deal of effort was concentrated during the early nineteen-forties into the methods of enabling the energy produced by the fission process to be released in large quantities and to be controlled for power production purposes.

It was known that in the act of fission of uranium two or three additional neutrons (fission neutrons) are released. If these neutrons in turn produce further fission, releasing more neutrons, then a fission chain reaction would result with the consequent release of a large amount of energy. Obviously, as neutrons in any particular system can be lost by absorption in non-fissile materials and uranium without causing fission or may escape completely from the system, then a chain reaction can only be sustained if the size of the system exceeds a certain critical value in which the number of fission neutrons produced is equal to or greater than the number lost by escape, absorption and other processes. A system in which a self-sustaining chain reaction is produced is a nuclear reactor.

The first nuclear reactor was built under the strictest security arrangements in the squash court under the University of Chicago's Stagg Field Stadium by Fermi and his collaborators.[3] The Chicago Pile (CP 1) diverged for the first time on the 2nd of December 1942, which was certainly an epochal event in the history of nuclear power.

Large numbers of reactors were constructed following the success of the Chicago Pile the main functions of which was the production of plutonium fuel for military purposes. However, in October 1956 a nuclear reactor constructed at Calder Hall in Cumberland fed electricity to the National Grid.[4] This was the first time that electricity had been produced on a commercial scale by nuclear means and heralded the new era of nuclear power.

Many nuclear power stations are now operational throughout the world and it appears that as a result of recent improvements in design that many more will be built in the future. The cost of electricity produced by the most economic stations at the present time is comparable to that produced by conventional stations and it is confidently expected that future stations will become more and more economical.

It is a remarkable achievement and one which reflects the energy and enthusiasm which has been directed into the nuclear engineering field that in the 25 years since the first reactor became operational over 500 nuclear reactors have been constructed.

1.2. AIMS OF REACTOR OPERATION

Several stages are necessary between the first concepts for a particular reactor system and the operation of the reactor at power. These are the initial feasability study, the design study, the construction and finally the initial fuel loading, commissioning and operation.

Before commencing a design study, several decisions of a fundamental nature have to be made. Firstly the type of reactor and its eventual use must be decided. Then consideration must be given to the type of fissile fuel to be used, the methods of containment or canning of the fuel, the type of moderator, the coolant, the method of control and type of control elements, the method of fuel loading and the movement and discharge of active fuel, the reactor shield, the nuclear and conventional instrumentation, the safety devices to be incorporated and the overall safety of the installation and finally the general overall design of the complete installation.

Reactors are used for a great many purposes. For instance, research reactors may be designed to provide extremely high neutron fluxes of up to 10^{15} n/cm² sec or even greater, or may be zero energy facilities with only limited application. They may, on the other hand, be designed for maximum flexibility with large numbers of experimental facilities. In the case of power reactors, the main concern is the attainment of high temperatures and the eventual production of electricity. Consequently initial design studies vary greatly in detail and in complexity dependent upon the type and use of the reactor under consideration. The United Kingdom has developed the natural uranium—graphite-moderated—gas-cooled reactor for its power programme. More emphasis has been placed in the U.S.A. on the use of enriched uranium water-moderated and cooled reactors of the boiling water or pressurised water type.

The operating staffs of a nuclear reactor are basically concerned with the initial fuel loading, commissioning and start-up of a newly constructed reactor and with the continual operation and maintenance of the reactor in a safe and efficient manner.[5] There should be considerable liaison between design and operating staff for a particular project. The operating staff have obviously to familiarise themselves with the detailed design of the reactor and its components and this is usually carried out during the constructional phase of the project. Operating personnel become more and more involved with the project and at the commencement of the initial testing of the conventional components they take over the administration completely. The remaining phases are then controlled by the operating staff who are fully responsible for setting the reactor and its components to work in accordance with the design concept.

The first object of the reactor operating staff is to plan an operational programme with a great deal of care. Due regard must be given to possible errors in the design data and all planned experimental procedures should be self-contained and no operation should be carried out which depends on the validity of any theoretical data. Throughout any project constant collaboration at each phase between the design and operating staffs is essential. The need for immediate consultation between designers and operators is required if some unforeseen result is obtained. During normal operation experimental information should always be fed back to the designers so that the most use is made of operating experience.

The possible hazards involved in the operation of the reactor system must be considered by the operating staff. A large number of safety devices are always incorporated in reactor systems. However, these cannot provide protection against every eventuality and therefore strict operational and administrative procedures must be followed to ensure maximum safety. Particular care should be given to operations involving fuel loading, criticality, power operation, radiation hazards, waste disposal and contamination.

Finally the operating staff are responsible for ensuring that there is no undue exposure of personnel to ionising radiation, that all work in radiation areas is strictly controlled and that the general public are not subjected to any radiation hazard whatsoever.

1.3. ADMINISTRATION AND ORGANISATION

In order to operate a nuclear reactor safely and efficiently an experienced and qualified operating staff must be available.[5] Obviously the number of staff required is very dependent upon the type of reactor and the number of additional facilities associated with the reactor. In the case of a small research reactor operated on a daily basis then the total staff required may only be three or four people. In the case of a large power reactor working on a shift basis a total staff of over 200 people may be necessary.

The typical operating organisation for a research reactor would consist of a reactor manager assisted by a reactor supervisor, reactor operators, maintenance staff and health physics staff. The reactor manager would have complete responsibility for safety and operation and the planning of all experimental procedures and programmes. The reactor supervisor would be responsible for routine operation, routine records and certification and the operating personnel and their duties. The number of reactor operators required depends entirely on the size and use of the particular facility. A minimum of two is considered desirable on any shift of 8 hours, hence a minimum of ten would be required to operate a five-cycle shift. The operators would be required to start up, operate and shut down the reactor and to carry out other procedures for a particular experiment. They would also be responsible for keeping routine records and a log book of all operations carried out. In addition they would be responsible for fuel loading and carrying out routine radiation surveys.

The maintenance staff would be responsible for ensuring that all reactor and fuel-handling components are working efficiently and safely. They would be responsible for carrying out a number of routine checks during operation to ensure that no unexpected component failure is likely to endanger the safety of operation. The number of staff required is again very dependent upon the size of the reactor in question. For small establishments it is usual for the operating staff to carry out maintenance duties during reactor shut-down periods hence reducing the number of people required considerably.

Health physics duties involve routine radiation measurements, ensuring that all personnel are adequately protected against hazards which may arise, keeping operational records of all measurements, arranging for the measurement of the radiation doses of operating personnel by film badges or other instruments and dealing with contamination and the care of personnel who may have received large radiation doses during any radiation incident. Again in small establishments, all routine measurements and records are carried out by the operating staff and a health physicist is appointed in an advisory capacity to the reactor manager. The reactor manager should assume responsibility for normal radiation protection procedures.

A typical operating staff organisation plan for a large power reactor is shown in Fig. 1.1.

1.4. SAFETY CONSIDERATIONS

Safety is of prime importance in the operation of reactors. During the design study of a reactor system a detailed safety assessment will be carried out. This assessment will consider the effects on the reactor of a number of postulated faults.[6] The operating staff should assess these predictions carefully and produce methods for dealing with any dangerous situation which may arise. Administrative procedures will be set up to deal with any foreseeable radiation hazard and with all fuel movement.

The responsibility for the safety of the reactor must lie with the reactor manager as all operations carried out will have been approved by him. However, it is usual to set up a safety committee for each installation, the function of which is to consider the safety aspects of all reactor operations and proposed experiments. The safety committee should consist of members with experience in the general field of nuclear

Reactor Operation

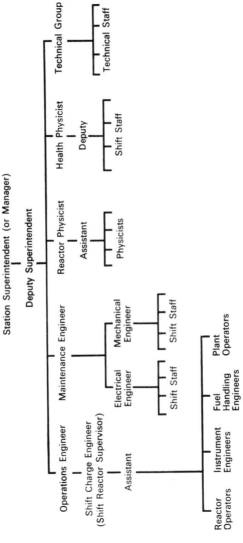

FIG. 1.1. Staff organisation for power reactors

safety and engineering although not necessarily familiar with the technical details of the particular reactor system. The reactor manager and the health physicist are usually members of the safety committee.

All planned procedures and experiments must be considered by the safety committee prior to being put into operation.

1.5. MAXIMUM CREDIBLE ACCIDENTS

For any new reactor installation a safety report is always produced in which the effects on the plant of faulty equipment and maloperation are considered in detail. The most important aspect of the safety report as far as reactor operation is concerned is the analysis of the effects of the maximum credible accident.[7,8]

The maximum credible accident is defined as the accident resulting from two simultaneously occurring failures of the reactor safety system coupled with incorrect remedial action by the operators. The reactor and its associated plant must be designed, if approval of the licensing authority is to be obtained, such that there can be no core melt-out or a widespread radiation hazard as a result of the maximum credible accident.

Consequently, as the performance of the plant may be limited by the criteria adopted for calculating the effects of the most feasible accident, elaborate calculations are usually carried out in order to prove the ability of the reactor equipment to deal effectively with most of the realistic faults which may arise.

The calculations are designed so that the dangers to the surrounding population and the operating personnel as a result of the maximum credible accident are effectively stated. The operating staff must be familiar with the expected occurrences following the accident and adequate procedures should specify the actions to be taken by all site personnel if the accident occurs.

Details of any particular maximum credible accident obviously depend upon the type, performance and characteristics of a reactor and only a brief discussion can be given here.

In small water reactors in which the reactivity worth of individual fuel elements is large in comparison with the total excess reactivity, the maximum credible accident is usually one involving the sudden rapid addition of reactivity to the core with the simultaneous failure of all safety devices.

In considering the effect of the sudden addition of reactivity, calculations must be carried out to determine:

(a) The initial rate of rise of neutron flux and power level.
(b) The temperature rise of the coolant and fuel.
(c) The effects of large temperatures on reactivity and the possible effects of void formation in the core due to coolant boiling.
(d) The thermal and mechanical stresses produced in the fuel and core components.
(e) The radiation levels produced in the vicinity of the reactor shield due to excessive reactor flux levels.

In the case of a gas-cooled graphite-moderated reactor in which fuel-loading accidents are not likely to cause appreciable reactivity addition to the core, the worst accidents are considered to be those involving complete loss of coolant with the reactor operating at high temperature.

The maximum credible accident is usually one which assumes that a bottom gas duct (coolant inlet duct) bursts with the simultaneous failure of all safety devices. Operating temperatures in a reactor are limited to ensure that in the event of a burst bottom duct occurring the probability of any fuel element catching fire and melting is less than 1 in 100. Hence, in the case of the nuclear power stations in the United Kingdom the calculated effects of the postulated maximum credible accident are used to limit the operating temperatures and hence the power output.

1.6. SITING REQUIREMENTS

Prior to the construction of a reactor installation consideration must be given to the choice of a suitable site.[9, 10] Sites are chosen with due regard to the release of radioactivity from the reactor following the occurrence of the maximum credible accident.[11] The ultimate concern is the possible exposure of the surrounding population to a major radiation hazard. Possible exposure paths for atmospheric release of radioactivity must be considered in detail. Radiation hazards following an atmospheric release may be due to:

(a) airborne contaminants which give rise to external radiation from air contamination, and internal radiation due to inhalation; and

(b) deposited contaminants which give rise to external radiation from contaminated buildings, clothes, ground, etc., and internal radiation due to ingestion of food and milk.

Many different practices are adopted in different countries for determining siting criteria. As there is no well-developed siting practice, there is a tendency for licensing authorities to adopt a very conservative point of view. The present situation is that both in the United States and the United Kingdom nuclear power reactors are built in fairly remote areas. However, the nuclear power industry has a tremendously good record regarding hazards to the general public. It is fairly true to say that if other industries took the same degree of care with regard to wastes then there would certainly be no river pollution or smog. In addition the records show that the total amount of activity released to the atmosphere from all nuclear power stations in the period 1960/63 was only 0·13% of the amount permitted by the regulations.

In view of these facts, licensing authorities are under some pressure to relax siting restrictions so that nuclear reactors can be constructed in more densely populated areas. It would certainly be possible to provide additional safety features which would adequately compensate for any additional risks in siting reactors in areas of high population density.

In establishing the suitability of any particular site, three distances are usually defined. These are:

(1) the exclusion area,
(2) the low population zone,
(3) the population-centre distance.

The exclusion area is the area immediately surrounding the reactor site which is controlled by the reactor operating staff. This area will be a fenced area with suitable notices indicating that a radiation hazard exists within the area. Access to the area is controlled by the operating authority.

The population-centre distance is the distance in miles from the reactor to the boundary of a densely populated area with a population exceeding 25,000 inhabitants.

The low population zone is an area surrounding the exclusion area in which the total number and density of the population must be such that there is a reasonable probability that there could be organised control following a radiation incident.

Different criteria are used in different countries for establishing the three specified zones. In the United States the zones are always specified in terms of a radiation dose whereas in the United Kingdom only the exclusion area is defined in this way. Other factors are taken into account to define the low population zone and the population-centre distance.

In the United States the exclusion area must be such that a person situated at the boundary for 2 hours immediately after the maximum credible accident has occurred would receive a dose less than 25 roentgens whole body or 300 roentgens to the thyroid. The low population zone must be large enough for a person situated at the boundary to receive a dose less than 25 roentgens whole body or 300 roentgens to the thyroid during the complete period of time when activity is released or may be released following the occurrence of the maximum credible accident.

In the United Kingdom the three distances must be such that following a maximum credible accident the radiation dose to the general public must be less than

(a) 20 roentgens external radiation for children and pregnant women and 30 roentgens for all others.

(b) 75 roentgens combined beta and gamma dose in any superficial tissue for children up to the age of 16 years and 150 roentgens for other persons.

(c) Internal radiation from inhalation and ingestion:
    Iodine—25 roentgens to the thyroid.
    Strontium-89—15 roentgens to the bone.
    Strontium-90—1·5 roentgens per year to the bone.
    Caesium-137—10 roentgens whole body.

In addition the radiation level at the boundary of the exclusion area during normal operation should be such that a person at the boundary shall not receive more than 0·5 rems in a year.

Many factors have to be considered before a final choice is made regarding any particular site. Some of the many problems to be tackled will be briefly outlined.

1.6.1. Reactor Type and Power Level

The suitability of any particular site is obviously very dependent on the type and characteristics of any reactor installation. Site selection will depend on the power level, the use of the reactor (i.e. research or power production), the number of experimental facilities, irradiation facilities, hot laboratories, the type of coolant, the amount of radioactive waste discharge (gaseous and liquid in particular), and the consequences of the occurrence of the maximum credible accident.

Power reactors have been given relative site ratings which make allowance for the degree of feasibility of a major accident. Gas-cooled reactors are subject to less severe accidents than water-cooled reactors and hence may be located on sites which are unsuitable for water-cooled reactors. Containment of a reactor will enable the site regulations to be fulfilled provided that the amount of release from the containment vessel can be well defined and is an acceptable amount.[12]

Research reactors are also classified with regard to site criteria. The reactors are divided into three classes. Class I includes all reactors with limited total excess reactivity which are inherently stable against excursions and which are safe against loss of coolant accidents. Swimming pool reactors and reactors with powers less than 100 kW are included in this category.

Class II reactors are those which are safe from loss of coolant accidents followed by a reactor shut-down and in which step reactivity changes are prevented by fuel-loading procedures and administrative control. This class includes reactors with powers less than 1 MW (i.e. Argonaut types).

Class III includes reactors which may be subject to core melt-out following a loss of coolant accident. Reactors in this category will have powers up to 3 MW (i.e. MTR reactor).

1.6.2. Land Costs

The cost of the land may have considerable influence on the choice of site. Although it is hoped that in the future it may be possible to build reactors near to more populated areas than has so far been customary, the relative land costs between town and country districts may be very restrictive.

1.6.3. Future Expansion

It is always advisable to choose a site which will enable the particular establishment to expand in the future. For instance, if a research centre is to be established, land must be available near to the reactor for hot laboratories, research laboratories and other facilities. All of these may wish to expand their fields of interest at some future stage.

It is common practice in the nuclear power industry to choose a site large enough for the eventual construction of more than one nuclear station.

1.6.4. Transportation

It is important to remember that good road and/or rail communication is highly desirable. The reactor must be constructed with ease of access to the site and provision should also be made for the removal of spent fuel from the site, which involves transportation of lead containers of 30 or 40 tons.

1.6.5. Use of Surrounding Land

The surrounding area will be in the low population zone, the implications of which have already been discussed. However it is highly undesirable to choose a site adjacent to chemical factories, oil refineries, public water works, hospitals and schools.

The density of the population must be carefully considered and the degree of control which could be put into operation in the event of an accident. In some respects factories are ideal neighbours as control can easily be exercised.

1.6.6. Floods
The possibility of flooding should be considered. Areas which are subject to regular and excessive flooding are usually avoided.

1.6.7. Meteorology and Geology
It is important to consider the meteorological and geological conditions pertinent to a site before selection is made.[13]

1.6.8. Cooling Water
The availability of large supplies of cooling water is often an important consideration. Cooling water is required for steam condensers in a power station and sites must be chosen near to large rivers, lakes or the sea.

1.6.9. Waste Disposal
Adequate means must be available for the storage and disposal of active waste. Particular importance must be attached to liquid and gaseous discharges.

1.6.10. Proximity of Manufacturing Firms
It is an economic advantage to choose a site which is near to the manufacturing industries concerned with the project.

1.6.11. Other Factors
Other factors which influence site selection include labour and building costs, living conditions near to the site, proximity of public water supplies, proximity of universities, interference with nearby installations and local building restrictions.

1.7. RULES AND LICENSING CONDITIONS

It is common practice in all countries for an appropriate government department to be responsible for authorising the construction and operation of a nuclear reactor by a private concern. The degree of participation of the government in the project varies from country to country. An appropriate authorisation is only granted when it has been demonstrated that the nuclear installation will be adequately and efficiently designed and operated.

In the United States the Atomic Energy Commission is the appropriate department concerned with the issue of various licences. Details associated with the granting of a licence are given in the Code of Federal Regulations (Federal Register). This consists of many sections each dealing with an appropriate part of a licence to operate an installation.[14]

In the United Kingdom the licensing authority is the Inspectorate of Nuclear Installations in the Ministry of Power. In order to construct and operate a nuclear reactor a licence is required under the Nuclear Installations Act 1965.[15] Any applicant who wishes to obtain a nuclear licence must present certain information to the inspectorate to enable an assessment to the made which will ensure that there is no public hazard.

Before a nuclear site licence is granted, the inspectorate must be given details of the proposed site, a general specification of the reactor and associated plant and a safety report defining the hazards that might arise in the event of a radiation incident. Each of these items will be subjected to a detailed assessment.

In the United Kingdom the following procedure is adopted when making an application to the Ministry of Power for a nuclear site licence.

1.7.1. **Assessment of the Site**

An applicant for a nuclear site licence must submit the following details in order that a proper appraisal of the site may be carried out:

1. The persons responsible for the design, construction, examination, inspection, testing, and operation of the reactor must be specified.[16]
2. The purpose of the installation and a brief description.
3. Wastes and proposed methods of disposal.
4. Exact site location, ownership and method of control of access.
5. Details of population and industrial installations in the vicinity of the site.
   (a) Population within radii of 220 yards, 440 yards and 1 mile for low-powered reactors and within $\frac{1}{2}$ mile, 1 mile and 5 miles for high-powered reactors.
   (b) Nearest towns with population of more than 1000, 10,000 and 50,000.
   (c) Utilisation of the surrounding area.
   (d) Public water reservoirs or works within 5 miles.
   (e) Surface drainage of catchment areas within 5 miles.
   (f) Foundations.
   (g) Prevailing meteorological conditions and any special weather characteristics.

The boundary of the nuclear site usually coincides with the exclusion area and a clearly marked fence is erected at this boundary to indicate that the area is a nuclear site and to prevent unauthorised access to the site. A condition of a nuclear licence is that proper security arrangements should be operative.

It is necessary at this stage in the application to define as precisely as possible the strength and quantity of all radioactive wastes which are likely to arise during the operation of the reactor. The proposals for dealing with active waste must be approved by the Inspectorate and Ministries of Agriculture, Fisheries and Food and of Housing and Local Government.

Finally, if the initial appraisal of the site is accepted by the Inspectorate, the applicant must then serve notice of his intention to construct a reactor on the local authorities, water boards and river board. Three months must elapse after the local authorities have been notified before a licence is granted to allow time for the local authorities to make representations to the Ministry of Power if desired. After this period a licence will be issued to commence preliminary constructional work on the project.

1.7.2. General Specification and Safety Assessment

The applicant must now present to the inspectorate the following detailed documents for his appraisal:

1. A general specification of the reactor and all associated components together with detailed electrical and mechanical drawings of all items.

2. A safety assessment of the reactor which must include a detailed statement of the expected behaviour of the plant for all possible faults and a complete treatment of the maximum credible accident.

3. Emergency procedures should be written giving courses of action required by all persons of authority in the event of a radiation incident, explosion or the outbreak of fire.

4. Operating instructions must be prepared to cover all aspects of reactor operation. These will include start-up, normal operation, shut-down, maintenance, and fuel-handling procedures and staff responsibilities.

5. A set of operating rules must be designed to safeguard the system and to set operating limits on all items of plant.

1.7.3. Final Approval for Operation

At the completion of the construction of the installation, proposals must be submitted to the Ministry stating in detail the tests to be carried out by the operators to prove that all items of plant are fully operational. On the satisfactory completion of these tests, proposals are submitted for a nuclear commissioning procedure commencing with fuel loading and ending with the initial raising of the reactor to power.

The results of the nuclear commissioning tests must be approved before clearance is given to commence normal operation.

REFERENCES

1. HAHN, O. and STRASSMANN, F. *Naturwissenschaften* **27**, 11 and 89 (1939).
2. MEITNER, L. and FRISCH, O. R. *Nature* **143**, 239 (1939).
3. Chicago Pile Number One. *U.S.A.E.C. Report* TID-292 (1955).
4. Calder Hall Atomic Power Station. *Nuclear Engineering* **1**, 7 (1956).
5. Manual for the operation of research reactors. *I.A.E.A. Technical Reports Series* No. 37 (1965).
6. Theoretical possibilities and consequences of major accidents in large nuclear power plants. *U.S.A.E.C. Report* WASH-740 (1957).
7. BURNETT, T. J. Reactors—hazard v. power level. *Nuclear Sci. and Eng.* **2**, 382 (1957).
8. NERTNEY, R. J. A standard practice guide for hazard analysis of experimental systems. *A.E.C. Research and Development Report* IDO 16466 (1958).
9. Guide for site selection criteria. *U.S.A. Federal Register* **26** (28), 1224 (1961).
10. Reactor site criteria. *Nuclear Safety* **5**, 3 (1964).
11. Regulation of radiation exposure by legislative means. *Nat. Bur. Standards Handbook* 61 (1955).
12. Containment of gas-cooled power reactors. *Nuclear Safety* **4**, 4 (1963).

13. Meteorological aspects of the safety and location of reactor plants· *World Met. Organisation Technical Note* No. 33 (1960).
14. Licencing of nuclear facilities. *U.S.A. Federal Register* **26** (197), 9653 (1961).
15. *Nuclear Installations Act 1965.* H.M. Stationery Office, London (1965).
16. Licencing of reactor operators. *U.S.A. Federal Register* **23** (130), 5064 (1958).

CHAPTER 2

Principles of Control and Operation

A NUCLEAR reactor, whether designed for power production or as a research facility, must have a control system which will enable the reactor and associated plant to be operated safely and efficiently.[1, 2, 3, 4] The control system must be capable of starting up the reactor, allowing the reactor to run at a steady continuous power level, allowing the power to be changed to any desired level and shutting down the reactor completely. In addition a number of safety devices must be included in the system, to shut down the reactor rapidly and automatically if certain faults occur or in the event of maloperation.

A high degree of reliability should be aimed at in the design of a control system. It is usual to provide a number of instruments duplicating certain control aspects so that the efficiency of the system is not impaired as a result of component faults or maintenance.

In order to provide power continuously the critical size of a nuclear reactor for a particular geometry must always be exceeded.

The critical size or critical loading is defined as the size or loading of the reactor which enables a self-sustaining neutron chain reaction to be achieved. At criticality the number of neutrons produced is equal to the number of neutrons lost by absorption and by loss from the surface of the geometrical arrangement. The effective multiplication factor (k_{eff}) is equal to unity at criticality. The amount by which criticality is exceeded is measured in terms of the reactivity of a system. Reactivity is equal to $(k_{eff}-1)/k_{eff}$, and if this is greater than unity then the system is said to possess excess reactivity. A negative reactivity value is a measure of the amount by which the reactor is subcritical.

There are three principal units of reactivity used in reactor physics. These are:

(a) *Percent reactivity*. This is the simplest unit and the one which will be used in this present text. The value of the reactivity of a system is expressed in terms of a percentage (e.g. if $k_{eff} = 1 \cdot 01$; then the excess reactivity of such a system will be $1 \cdot 0\%$).

(b) *Dollars and cents*. In this case the reactivity scale between 0 and β (the fraction of delayed neutrons compared with all fission neutrons—see equation (6.7)) is divided into units of 100 cents (or 1 dollar). In a U-235 reactor the value of β is $0 \cdot 0064$ and therefore a reactivity of $0 \cdot 64\%$ corresponds to 1 dollar. Hence to convert from dollars to percent the reactivity value in dollars is multiplied by β.

(c) *Inhour:* This unit is not used so frequently as the previous ones. It is defined such that a reactivity of one inhour corresponds to a stable reactor period (*e*-folding time) of 1 hour.

For long periods of operation at high temperatures and large flux levels, reactivity in excess of the critical value ($k_{eff} = 1$) must be built into the core to overcome the negative reactivity effects of temperature, fuel poisons and absorbing materials in the core. In the case of a research reactor an additional excess reactivity of the order of 1 to 2% ($k_{eff} = 1 \cdot 01$ to $1 \cdot 02$) may be required, whereas for a high-temperature power reactor as much as 5% ($k_{eff} = 1.05$) may be needed. Consequently at the initial start-up of a reactor a large amount of excess reactivity is available and the fission rate and neutron flux could be capable of increasing extremely rapidly. For this reason the control system should be capable of preventing a large release of reactivity during the initial periods of operation.

Control of a reactor is usually carried out by varying the effective multiplication factor. If the value of k_{eff} is held at unity, then the power level of the core will remain at a steady level, if k_{eff} is increased above unity the power level will rise and if k_{eff} is reduced below unity, the power will fall and the reactor will be shut down.

2.1. METHODS OF CONTROL

The effective multiplication factor of a given reactor system may be changed by addition or removal of:

1. Nuclear fuel.
2. Moderator material.
3. Reflector material.
4. Coolant (to alter core temperatures).
5. Neutron absorbers.

Each of these methods or a combination of the methods has been used in practice.

Methods 4 and 5 are more suitable for the control of thermal reactors in which the fission neutrons are moderated to thermal energies before causing fissions in the reactor fuel. Thermal neutron absorption cross-sections are extremely high for some materials. In the case of fast reactors the fission neutrons themselves cause fission in the reactor fuel and such a system has no moderator. In this case control is achieved by the withdrawal and insertion of parts of the core (i.e. the fuel or reflector). Hence methods 1 and 3 are more suitable for fast reactors. With thermal systems reactivity is increased by the withdrawal of the neutron absorbers, whereas with fast reactors reactivity is increased by the insertion of fuel or reflector to the core. In most thermal reactors the reactivity is controlled by the withdrawal and insertion of a number of neutron absorbers. The type of system adopted and the actual number of the absorbers used is dependent on the design and characteristics of a particular reactor. However, there are certain general requirements which are common to all installations. The absorbers are usually constructed in the form of flat plates or cylindrical rods containing boron or cadmium sheathed in materials which are compatible with the materials within the reactor core. Boron and cadmium are used as the absorber on account of their extremely large neutron absorption cross-sections (i.e. boron-10—4,000 barns and cadmium-113—25,000 barns). Aluminium and stainless steel are usually used as sheathing materials.

Mechanisms must be provided for withdrawing and inserting the control absorbers in the core. The absorbers are usually moved vertically in and out of the core and are attached to the driving mechanisms by a variety of methods. It is desirable to arrange for the driving mechanisms to be situated outside the core and the shield in order to provide ease of access for maintenance purposes. Important considerations for the driving mechanisms are:

(1) Simple design to give the maximum reliability.
(2) Provision of a device (e.g. magnetic clutch) to enable the absorbers to be rapidly inserted into the core.
(3) Provision for the complete removal of the mechanism and absorber from the core.

(4) Constant driving speeds for insertion and withdrawal.

(5) Rates of withdrawal to be limited to prevent rapid reactivity addition to the core.

The control system will be operated by an operator or by an automatic device as a result of information supplied by neutron flux measuring instruments, temperature recorders and other reactor instrumentation.[5] The overall control action is represented schematically in Fig. 2.1.

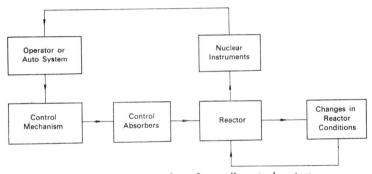

FIG. 2.1. Representation of overall control system

The overall control system should be designed such that:

1. The range of the control system will cover complete operation.

2. Provision should be made for sufficient fine adjustment of the control parameters.

3. There should be quick response to all activating signals without causing undue disturbances.

4. The system must be designed to fail-safe if an emergency occurs or a component is faulty.

2.2. RANGE OF CONTROL

The system of control absorbers used for any particular reactor must be capable of covering the complete reactivity range.[5, 6] The following factors must be considered in establishing the amount of excess reactivity to be built into a reactor core.

2.2.1. Temperature

Most reactors have a negative temperature coefficient of reactivity (i.e. reactivity change per °C rise in temperature).

Hence, the excess reactivity will decrease as the temperature rises and a reactor which is just critical at ambient temperature will become subcritical as the temperature rises. Therefore an additional amount of excess reactivity (fuel above the critical size) will have to be added to the reactor if criticality is to be maintained at high temperature. Typical reactivity effects due to temperature are:

(a) For an enriched-uranium water reactor: Temperature coefficient $= -6 \cdot 0 \times 10^{-5}$ per °C. Hence reactivity change due to a 20°C rise in temperature would be $-0 \cdot 12\%$.

(b) For a graphite moderated-natural uranium reactor: Temperature coefficient due to fuel $= -2 \cdot 0 \times 10^{-5}$ per °C. Temperature coefficient due to graphite $= -3 \cdot 0 \times 10^{-5}$ per °C (at zero irradiation). Hence for a rise of 200°C, a reactivity change of $-1 \cdot 0\%$ would be produced.

2.2.2. Fuel Depletion

The fissile material in the reactor core will gradually be used up during the life of the reactor. Consequently in order to avoid having to replenish the fuel at a very early stage during operation it is usual to build in additional reactivity to allow for this fuel burn-up or fuel depletion. The amount of additional reactivity required is dependent upon the fuel burn-up rate.

A typical value for a power reactor operating on a 2000 MWD per Tonne cycle would be -0.2% in reactivity.

2.2.3. Poisoning

A number of nuclei are produced as a result of the fission process which have large neutron absorption cross-sections. These reduce the reactivity of the core due to additional loss of neutrons by absorption. The most important of these poisons during operation are xenon-135 with an absorption cross-section of 3.5×10^6 barns and samarium-149 with a cross-section of 53,000 barns (see Chapter 9).

Xenon is produced as a result of the radioactive decay of the fission product tellurium which decays with the chain

$$\text{Te-135} \xrightarrow[2 \text{ min}]{\beta^-} \text{I-135} \xrightarrow[6.7 \text{ hr}]{\beta^-} \text{Xe-135} \xrightarrow[9.2 \text{ hr}]{\beta^-} \text{C}_s\text{-135} \xrightarrow[2 \times 10^4 \text{ yr}]{\beta^-}$$
$$\text{Ba-135 (Stable)}$$

On account of the time constants associated with production and decay of the xenon-135 isotope the effect of xenon on reactor operation is restricted to periods of several hours. The total reactivity change produced as a result of the build up of xenon to an equilibrium concentration depends on the level of the neutron flux during operation. The effect is small for fluxes less than 10^{12} n/cm^2 sec, at 10^{12} the reactivity effect is approximately -0.7%, at 10^{13} is -2.0 to -3.0% and a limiting value of -4.8% is reached at fluxes in excess of 10^{15} n/cm^2 sec.

Samarium is produced as a result of the decay of the fission product neodymium with the decay chain

$$\text{Nd-149} \xrightarrow[1.7 \text{ hr}]{\beta^-} \text{Pm-149} \xrightarrow[47 \text{ hr}]{\beta^-} \text{Sm-149 (Stable)}$$

The equilibrium value for the reactivity effect of samarium-149 is $-1 \cdot 2\%$ which is independent of the neutron flux level in the reactor (see Chapter 9).

2.2.4. Experiments and Absorbers

Additional reactivity must be built into the reactor core to compensate for any absorbing effects of the experimental facilities. Obviously much more would be required if complex in-core experimental rigs are to be provided and in this case the reactivity range to be covered could be much more extensive.

2.2.5. Control Rods

With the reactor operating at temperature and all poisons fully built-up additional reactivity must be available in order to provide efficient control. This reactivity will be absorbed by a number of partially inserted control rods. In general $0 \cdot 2$ to $0 \cdot 3\%$ in reactivity is considered to be adequate for this purpose.

The reactivity range required for typical research and power reactors is summarised in Table 2.1.

TABLE 2.1. REACTIVITY RANGES FOR REACTORS

Effect due to	Negative reactivity $\frac{\Delta k}{k}$ effective $\%$	
	Low-power reactor	High-power reactor
Temperature	0·12	1·40
Fuel depletion	—	0·20
Poisoning (equilibrium)	—	2·00
Experiments and absorbers	0·20	0·20
Control	0·20	0·20
Total	0·52	4·00

2.2.6. Reactivity Safety Margin

The control system for any reactor must be capable of shutting down the reactor completely. Consequently the control absorbers must be capable of absorbing all the built-in excess reactivity plus an additional amount to reduce the effective multiplication factor below unity when all the absorbers are fully inserted. The negative value of the reactivity of a reactor with all control absorbers fully inserted is known as the shut-down capacity of the core. The value of the shut-down capacity is the difference between the total excess reactivity of the reactor in the cold, unpoisoned state and the total reactivity worth (i.e. negative reactivity effect) of all the control absorbers. In the case of research reactors very rapid shut-down is often required, especially if transient experiments are being carried out and consequently a large shut-down capacity of the order of 5 to 10% is required. In some cases the water moderator is completely dumped from the core at shut-down. The effect of this is to provide a shut-down capacity of the order of 30% in reactivity.

For power reactors a shut-down capacity of the order of 1·5 to 2·0% is considered an adequate safety margin.

2.3. EFFECTIVENESS OF CONTROL ABSORBERS

At the design stage calculations are carried out to determine the reactivity worths (i.e. reactivity absorbed) of the control absorbers used in the reactor. In view of the difficulty in obtaining reliable theoretical information a number of experiments are carried out during the initial loading of fuel into the reactor core and with the reactor fully loaded to determine the reactivity absorbed by the control system.

The variation of the amount of reactivity absorbed by an individual control absorber with depth of insertion in the core is shown in Fig. 2.2. This shows the characteristic S-shaped curve for the variation of reactivity with position in the core. From the control point of view it is important to realise that the maximum effect per unit length of the absorber occurs at

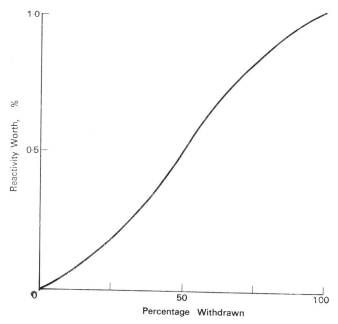

FIG. 2.2. Variation of control rod worth with depth of insertion

the centre of the core at the position of maximum neutron flux. At the ends of the travel of the rod (i.e. almost fully inserted and almost fully withdrawn) the reactivity change per unit length is very small. This effect is shown in Fig. 2.3 which is the differential plot of the worth of a unit length of the control rod at different depths of insertion.

Rapid reactivity changes cannot be produced if the controlling rods are near the ends of their travel and it is good operating practice to make sure that controlling rods are operated at the most effective depth of insertion.

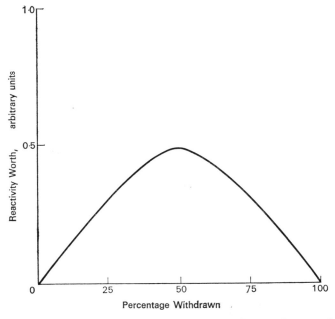

FIG. 2.3. Differential control rod worth for unit length of a control rod

The effectiveness of a control rod in a given reactor depends upon its location in the core.[5] Since the neutron flux is a maximum at the centre of the core then a rod placed in the centre will be more effective than at any other position across the core. The effectiveness of a rod at any point in a cylindrical reactor is found to be proportional to the square of the thermal neutron flux at that point. Hence in the case of a cosine flux distribution the worth of a rod located halfway along the core radius would be approximately half the worth of a central rod.

It is usual for a power reactor to have large numbers of control rods. In this case it is found that interference effects

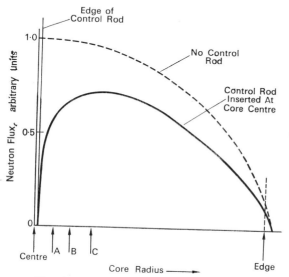

FIG. 2.4. The effect of inserting control rods

are produced between the rods such that the total worth of a number of control rods is less than the sum of the individual worths. This interference effect is dependent upon the relative location of the rods with respect to each other and the spacing between them. It is found that as the spacing between two rods is increased the total reactivity worth increases to a maximum and then decreases with further increase in the separation distance. These interference effects result from the distortion of the reactor flux distribution by the presence of a control absorber. The effect of rod insertion on the flux distribution is illustrated in Fig. 2.4. The dotted curve is the original flux distribution with no control rod. The full curve shows the flux distribution after inserting a rod at the centre of the core. Now if a second rod is inserted at A the total worth will obviously be less than the individual worths. If the rod is inserted at B the total worth will be greater than at A as the second rod has been placed in a position of higher flux. The total worth will increase to a maximum at position C, and then decrease.

2.4. FLUX SHAPING

In the case of a large power reactor a large number of control rods may be required to cover the available reactivity range. The rods should be distributed in a symmetrical way such that a uniform flux distribution is maintained as far as possible during normal operation. In addition every attempt should be made to withdraw rods in groups so that large flux distribution distortions do not occur. Large fluctuations in the coolant outlet temperatures will be produced if large flux distortions occur during rod movement.

A requirement of a power reactor is to produce the maximum possible coolant outlet temperature which will be achieved by obtaining the maximum outlet temperature from each individual fuel channel. The maximum channel outlet temperature is produced in channels in positions of maximum neutron flux which occur at the centre of reactors. The channel outlet temperature may be increased in a large number of channels by producing a flat flux distribution. Control absorbers are used to flatten the neutron flux in the central region of the reactor hence increasing the total heat output of the reactor.

Figure 2.5 shows how flux flattening is achieved by inserting absorbers in the central region to shape the neutron flux. Obviously if flux flattening is carried out then additional reactivity must be built into the reactor to compensate for the additional absorbers placed in the core. Flux flattening and

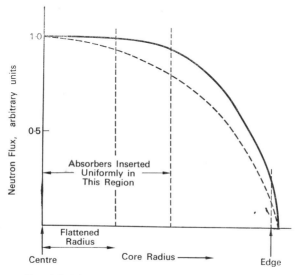

FIG. 2.5. Flux flattening by insertion of absorbers

shaping is always carried out by trial and error methods on an operating reactor. Rods are inserted and withdrawn at different positions and measurements made of the flux distribution and the channel temperatures. The procedures are repeated until satisfactory flux and temperature distributions have been obtained.

2.5. TYPES OF CONTROL RODS

The total reactivity range to be controlled by a control rod system of a power reactor may be as much as 6.0%. The control rods must be able to control the total excess reactivity (of order 4.0%) and provide an adequate shut-down margin (of order 2.0%). In view of this large reactivity range, different rods are used for different functions. The same applies in research reactors although in this case the total excess reactivity may be of the order of 0.5% with a shut-down capacity of order 3.5% (i.e. total reactivity range of 4.0%).

Control rods are divided into a number of groups each having a different function. These are:

1. Safety rods—normally held fully withdrawn providing additional negative reactivity in the event of an emergency. They control a large amount of reactivity and have slow withdrawal speeds.
2. Main control rods—used for raising the reactor power from shut-down to full power. They control a large amount of reactivity and have slow withdrawal speeds.
3. Fine control rods—used for controlling the reactor power level during normal operation. They control a small amount of reactivity and have fast withdrawal speeds.
4. Flux shaping rods—used for shaping and flattening the neutron flux.
5. Automatic rods—used for automatic control or for control of reactor instabilities.

A small research reactor controlling an excess reactivity of 1.5% could have four control rods with the following functions:

(a) One safety rod of worth 1·5%.

(b) Two main control rods of total worth 3·0%.

(c) One fine control rod of worth 0·5%.

Total shut-down capacity with this arrangement would be 3·5%.

A large power reactor may have a large number of control rods. Berkeley Power Station has a total of 132 rods with the functions given in Table 2.2.

With a total excess reactivity of 4·5%, the arrangement shown in Table 2.2 will give a shut-down capacity of 5·5%.

TABLE 2.2. FUNCTIONS OF CONTROL RODS FOR LARGE POWER REACTORS

Type of rod	Total number	Reactivity controlled	Function
Safety group	20	1·0%	Held withdrawn for emergency use
Group A	56	4·5%	Shut-down rods. Raise power from shut-down to criticality
Group B	30	2·3%	Operating rods. Raise power from criticality to maximum
Group C	8	0·7%	Used for flux flattening and shaping
Automatic group	18	1·5%	Automatic control of instabilities

2.6. THE EFFECTS OF CONTROL ROD
WITHDRAWAL

On withdrawing the control rods of a reactor a certain amount of excess reactivity will be released. This will be associated with a corresponding increase in neutron flux level in the core due to increased multiplication.

On exceeding criticality (i.e. $k_{eff} > 1$) then the reactor flux will increase, this increase being governed by the total amount of excess reactivity which has been released (i.e. the amount by which k_{eff} exceeds unity).

For a given excess reactivity the reactor period (the time for the reactor flux to increase by a factor e) is given by the inhour equation (equation (6.7), Chapter 6). This period is dependent on the numbers of delayed neutrons produced in the fission process. In fact control of reactors would be extremely complex if it were not for the influence of the delayed neutrons.

In the fission process 99·24% of all the neutrons emitted occur as prompt neutrons and are emitted within 10^{-14} sec of fission occurring. However, neutrons are also produced as a result of the decay of certain fission products and are therefore released at certain periods after fission has occurred. These neutrons are dependent upon the half-life of the radioactive fission products and are known as delayed neutrons. Their yield is approximately 0·76% of all neutrons emitted during fission. There are six main groups of delayed neutrons with varying percentage yields as given in Table 2.3.

These have been identified with the decay of certain fission products. For instance, groups 1 and 2 are a result of the radioactive decay of bromine-87 and iodine-137.

Although the total percentage of the delayed neutrons produced in fission is only 0·76% they have a considerable in-

TABLE 2.3. DELAYED NEUTRON GROUPS

Group	Half-life, sec	% yield
1	55·6	0·025
2	22·0	0·167
3	4·51	0·214
4	1·52	0·243
5	0·43	0·086
6	0·05	0·025

fluence on the rate of rise of the neutron flux with reactivity. Figures 2.6 and 2.7 indicate the effect of delayed neutrons on the rate of rise of neutron flux for a step change in reactivity of 0·1% and a ramp change of 0·001%/sec. These show clearly how easy it is, as a result of the presence of delayed neutrons,

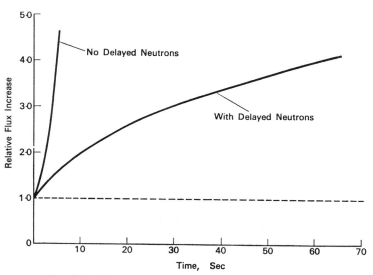

FIG. 2.6. Effect of a step change of 0·1% in reactivity

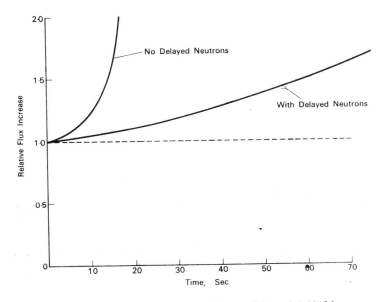

FIG. 2.7. Effect of a ramp change in reactivity of 0·001% per second

to control the rate of rise of the neutron flux by a control rod system.

An important point to note is that when the excess reactivity of the reactor is equivalent to the delayed neutron fraction (i.e. 0·76%) then the reactor will be critical on prompt neutrons alone and delayed neutrons are not necessary to sustain the chain reaction. Such a condition must be avoided during operation as the safety devices will not be capable of operating rapidly enough to control the rise of flux and a reactivity excursion would be produced. This condition is avoided by the basic design of the reactor control system.[7, 8]

2.7. FUNCTIONS OF OPERATORS

One of the most important aspects of efficient reactor operation is the suitable choice of instruments to supply data to an operator at a control desk. A good control system will be designed to assist the operator in every way. Sufficient information for safe and efficient control must be available. If too much information is presented then this will lead to confusion and a loss of efficiency. The operator may be assisted by the provision of a number of automatic devices. For instance, certain emergency shut-down systems must be provided as they can act more quickly than an operator in an emergency. It is also extremely useful to provide an automatic system for operating the reactor at steady power for long continuous runs. This will relieve the operator from routine adjustments and leave him free to keep an overall watch on reactor conditions. However, for start-up and non-routine operations the operator must take complete control. The basic duties of the operator are to obtain information and to exert controlling action as a result of this information. The operator will use his experience to reach decisions about the controlling action required for any particular set of reactor conditions.

2.8. LAYOUT OF CONTROL DESKS

The following information required by an operator is displayed at a control desk:

1. Reactor power and neutron flux.
2. Alarm indications for faults requiring immediate action.
3. Emergency faults.
4. Control rod positions and operating controls.
5. Other essential information.

In most control rooms the display instruments are divided into primary controls on a control desk and secondary controls on racks of instruments placed around the walls of the control room. The primary controls are those which require the particular attention and action of the operator. All other less important instruments are arranged in the secondary group of controls. A typical control desk for a small research reactor is shown in Fig. 2.8.

Instruments for a power reactor would be arranged in the following general manner.

2.8.1. **Primary Controls**
1. Control rod positions (rods used for fine control only).
2. Control rod selector switches and operation switches.
3. Linear power indicators and recorders.
4. Backed off power indicators and recorder.
5. Doubling time meters (or period meters).
6. Fuel element temperature recorders (hottest fuel elements).
7. Coolant temperature recorders (hottest coolant outlet temperature and inlet temperature).
8. Reactor coolant flow control indicators, recorders and operation switches.
9. Logarithmic indicator and recorder associated with start-up channels.
10. Emergency scram button.

2.8.2. **Secondary Controls**
1. Alarms for warnings and emergencies.
2. Shut-down indications and shut-down amplifiers.
3. Coolant flow recorders for heat exchangers and other parts of coolant circuit.
4. Main logarithmic power recorder.

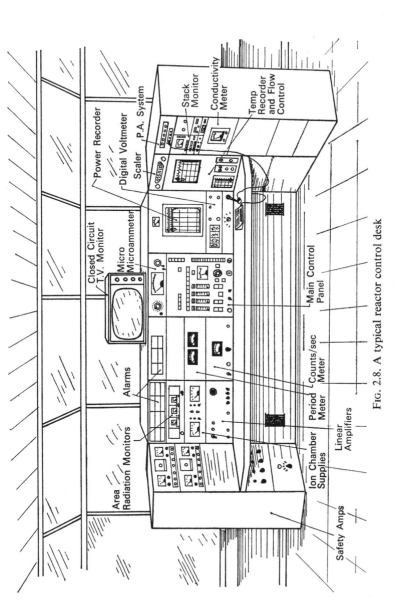

Fig. 2.8. A typical reactor control desk

5. Fixed control rod positions and positions of flux shaping rods.
6. Heat exchanger conditions (coolant, water and steam temperature and flow information).
7. Reactor coolant bulk inlet and outlet temperatures.
8. Coolant pressure.
9. Coolant circulators (operating information).
10. Autocontrol system supplies and information.
11. Burst fuel element detection gear information.
12. Turbine house information.
13. Electrical switchgear information.

2.9. WARNINGS AND EMERGENCIES

A number of conditions arise during normal operation which can effect the safety of the reactor and the plant. Consequently a warning system is incorporated into the reactor control instrumentation to indicate to the operator that a fault has arisen or to shut down the reactor automatically if a dangerous situation occurs. The aim of the warning and automatic shut-down system is to provide the maximum safety of operation and every attempt is made to design the system to give adequate protection against fault conditions without inhibiting the efficient operation of the reactor.[9, 10]

2.9.1. Emergency Shut-down

Conditions which require automatic shut-down depend to some extent on the characteristics of a reactor and on the results of the safety analysis of the faults which may arise.

In general automatic protection is provided for the following conditions:

1. Excess power level
    (usually set to operate at 10% to 50% above the operating level).

2. Doubling time or reactor period less than 5 sec.
3. Pre-set low power level
   (set at any desired level and used to provide protection at start-up and during low power operation).
4. Manual
   (shut-down button on control desk. In some cases other emergency shut-down buttons are placed near important locations around the reactor).
5. High or low coolant flow.
6. Excessive reactor temperatures.
7. Beam holes or shielding doors open.

In most reactor control systems some form of electromagnetic clutch is installed between the control rod mechanism driving motor and the rod withdrawal mechanism. In the event of the power supply to the electromagnetic clutch being cut off then the driving motor is disconnected and the control rod is free to fall. In some cases the free fall of the rod is assisted either by attaching springs to the control rod mechanisms or by using power-assisted units. The supply to the electromagnetic clutch or other device is operated by a relay system when a fault occurs. It is a feature of the safety system to provide a rapid insertion of the control rods following an emergency shut-down. The actual scram period permitted is dependent upon the reactor characteristics and varies between a few milliseconds for transient reactors to approximately 5 sec for a gas-cooled natural uranium reactor.

Following an emergency shut-down, the reason for the fault must be established before reactor start-up may recommence. In addition it often takes a considerable period of time to start up a large power station from complete shut-down (of order 24 hr in some cases). Consequently it is usual to operate certain shut-down devices in two out of three operation to prevent as far as possible shut-downs being produced as a result of spurious signals in the instrumentation. Also it is essential to limit the number of faults causing complete shut-down to those which immediately give rise to dangerous and unsafe reactor conditions.

2.9.2. Warnings

A large number of faults may arise during reactor operation which, if not rectified immediately, could produce unsafe conditions. However, to prevent the unnecessary shut-down of the reactor, these conditions give a warning indication to the operator by visual and audible means so that his attention is immediately drawn to the fault and corrective action taken quickly.

Conditions which fall into this category are:

1. Excess power level within 10 to 15% of the trip level.
2. Doubling time or period less than 10 sec (if the trip level is 5 sec).
3. Loss of power to the measuring instruments.
4. Loss of power to the safety channel instruments.
5. High radiation levels at certain locations.
6. Loss of an interlock condition.
7. Unusual reactor temperatures.
8. Coolant flow or temperature faults.
9. High counts on burst fuel element detection equipment.
10. Heat exchanger faults.
11. Faults associated with other items of plant which effect reactor conditions.

Table 2.4 lists the warnings and shut-down conditions associated with a typical gas-cooled, graphite-moderated natural uranium reactor.

TABLE 2.4. WARNINGS AND EMERGENCIES FOR GAS-COOLED POWER REACTOR (MAXIMUM POWER 550 MW)

Condition	Operation	Trips	Warnings
Start-up channel	1 out of 2	50 W power 20 sec period	—
Low log power channel	1 out of 2	10 MW power 20 sec period	5 MW power
Main log power channel	1 out of 1	20 sec period	Variable 1 to 600 MW
Excess power (shut-down amplifier channels)	2 out of 3	Variable up to 600 MW	(1) At 20 MW below trip (2) At 100 MW below trip
Fuel element can temperature	12 units in 3 groups of 4 operated as 2 out of 3 groups	Variable 400 to 500°C for normal temperature of 420°C; trip set at 440°C	At 10°C below trip level

(continued overleaf)

TABLE 2.4 *(continued)*

Condition	Operation	Trips	Warnings
Duct gas outlet temperature	2 units per duct operated on 2 ducts out of 8 or 1 out of 4	Variable 200 to 450°C for normal temperature of 345°C; trip set at 365°C	At 15°C below trip level
Rate of change of gas pressure	1 out of 2	10 psi per minute	—
Coolant circulator failure	2 circulators out of 4	Loss of circulator	—
Manual	1 out of 2	Shut-down button	—
Burst fuel element detection gear	—	—	High count rate at pre-determined value
Temperature recorders	—	—	At certain temperature values
Area radiation monitors	—	—	Measured levels above pre-determined value

2.10. INTERLOCKS

In order to be able to start up a reactor in a safe and reliable manner certain conditions must be satisfied before control rod withdrawal commences. In general terms all auxiliary equipment must be functional, sources installed, nuclear instrumentation fully operational, safety devices operative, coolant flow and temperature conditions satisfactory for initial start-up, etc. Start-up procedures are always available to cover the correct sequence of events. However, in order to achieve reliability and maximum safety, it is usual to incorporate in a control system a number of relays which are interlocked so that a desired start-up sequence must be carried out. The interlocks prevent power being supplied to the electromagnetic clutches in the rod mechanisms until all desired conditions are fulfilled.

2.11. SAFETY LINES

It is normal practice to adopt the primary and secondary guard line system for the emergency shut-down circuits. In this system each action resulting in an emergency shut-down provides a break in two separate circuits. Therefore each trip condition has a back-up or secondary protective circuit which provides additional safety. Figure 2.9 shows a typical arrangement of the primary and secondary system normally adopted.

The four main relays A, B, C and D have contacts in the lines providing power to the electromagnetic clutches in the control rod mechanisms. Hence for power to be supplied to withdraw the control rods, these four relays must be energised. In the event of a period trip occurring, contacts P1 in the primary line and P2 in the secondary line will open. Hence relays A and B will be de-energised and contacts A1 and B1 in the primary line and A2 and B2 in the clutch supply line will open thus tripping the control rods. The secondary line provides back-up to the action because if P2 opens, relay F is de-energised and hence contact F2 opens de-energising relays C and D and opening contacts C2 and D2 in the rod clutch supply lines.

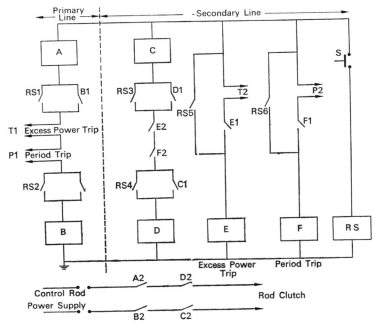

Fig. 2.9. Typical arrangement of primary and secondary guard line system

An interlock system is provided so that if the fault in the period meter circuit is immediately rectified without any corrective action, then closing contacts P1 and P2 cannot automatically restore power to the rod drive system. In this case P1 and P2 close, but A2, B2, C2 and D2 remain open as relays A, B, C, D and F are still de-energised. The operator must close the re-set contact S in order to energise the relays again. If S is closed, relay RS is energised, and hence contacts RS1, RS2, RS3, RS4, RS5 and RS6 close energising relays A, B, C, D and F and so restoring the supply to the clutch in the rod drive system. This interlock system ensures that following the rectification of the fault which produces the trip the operator must carry out an additional action before the safety lines are again operational. In all reactor control and interlock circuits the relays are installed in the normally energised state so that they fail-safe in the event of a loss of supply.[11]

REFERENCES

1. BINNS, J. E. Design of safety systems for nuclear reactors. *Nuclear Safety* **4,** 2 (1962).
2. SCHULTZ, M. A. Reactor safety instrumentation. *Nuclear Safety* **4,** 2 (1962).
3. MOORE, R. V. *Proc. Inst. Elec. Eng.* **100,** 1, 90 (1953).
4. SCHULTZ, M. A. *Control of Nuclear Reactors and Power Plants.* 2nd ed. McGraw-Hill (1961).
5. GLASSTONE, S. and SESONSKE, A. *Nuclear Reactor Engineering.* Van Nostrand (1963).
6. HARMER, J. M. *Nuclear Reactor Control Engineering.* Van Nostrand (1963).
7. COX, R. J. and SANDIFORD, K. R. Reactor control and instrumentation. *J. Brit. Nuc. Eng. Conf.* **2,** 2 (1957).
8. COX, R. J. and WALKER, J. The control of nuclear reactors. *J. Brit. Nuc. Eng. Conf.* **1,** 106 (1956).
9. Design of safety systems for nuclear reactors. *Nuclear Safety* **4,** 2 (1962).
10. SIDDALL, E. Reliability of reactor control systems. *Nuclear Safety* **4,** 4 (1963).
11. JACOBS, I. M. Safety system design technology. *Nuclear Safety* **6,** 3 (1965).

CHAPTER 3

Nuclear Instrumentation

3.1. GENERAL PRINCIPLES

In order to provide adequate control of nuclear reactors at all stages of operation information concerning reactor conditions must be obtained by means of suitable instrumentation.[1, 2, 3] For instance, the operator should have a knowledge of the state of the neutron flux in the core, the rate of reactivity change, the power generated and detailed information of core temperatures. In addition it is often useful to obtain details of any local disturbances which may occur in the core.

Power level, reactivity and reactor temperatures are due to the fission process which is proportional to the neutron flux in the core. Operating information may be obtained by measuring any one of these parameters. In a large power station information about neutron fluxes in the core is obtained by detailed temperature distribution measurements. Temperature changes are due to flux changes and by the suitable positioning of thermocouples extremely accurate and detailed information may be obtained.

However, during reactor start-up, fuel loading and before temperatures have increased other methods must be adopted. In the case of many research reactors there is inadequate temperature monitoring or only low temperatures are produced and therefore other methods must again be employed to provide operating information. The required information i nearly always obtained by using instruments to measure the neutron flux produced in the core, near to the core or in an external thermal column.

3.2. RANGE OF MEASUREMENT

One of the main problems associated with the provision of suitable nuclear instrumentation is the coverage of the operating range over which measurements are required.[2, 3, 4] For instance, the reactivity of a shut-down reactor is normally between -2.0% and -10.0% and the power generated is of the order of a few milliwatts. During operation the power generated may be hundreds of megawatts and consequently it may be necessary to cover a range of ten or eleven decades in neutron flux. This is far beyond the range of any one type of instrument and it is usual to use several instruments with overlapping ranges to cope with the situation.

It is extremely important to provide instruments to cover the complete flux range from shut-down to full power. An instrument must always indicate a change in flux when a control rod is moved or if any other change takes place in the core. If the lower end of the range is not covered adequately then during start-up the operator would have to withdraw control rods carefully and allow the flux level to rise until it reached the operating range of the instruments. Such a situation is known as a blind start-up and can easily lead to a dangerous situation. Due to some unknown fault a larger amount of reactivity may be released during rod withdrawal than anticipated and by the time the flux was indicated a short doubling time could be achieved which could lead to an excursion before corrective action could be taken. Hence any form of blind start-up cannot be accepted and methods must be used to overcome this problem.

3.3. POSITIONING OF INSTRUMENTS

Most detectors are more sensitive to thermal neutrons as they depend upon reactions such as the thermal neutron-alpha particle reaction with boron or the fission reaction with uranium. The alpha particle (from the reaction with boron) and fission products (from the fission of uranium) are then used in the detector to ionise a gas and produce an ionisation current. Consequently for maximum efficiency of a system the detectors should be placed in a thermal neutron flux. In the case of research reactors the instruments are usually placed adjacent to the core reflector for low power operation. At high powers provision is made to withdraw the instruments into holes in the shield or into a thermal column (to provide thermal neutrons) as the power rises. For power reactors it is a disadvantage to place the instruments near to the core as:

(a) The life of the detector is limited if placed in a high flux or high temperature.

(b) The detector and associated cables will become radio-active.

(c) Local discontinuities in flux near to the instruments will produce readings unrelated to the normal power level.

It is more usual to place the instruments in a thermal column positioned some distance away from the core; for instance outside the reactor pressure vessel. Leakage flux from the core is measured in this case and an efficient thermal column is required as the leakage flux has a high proportion of fast neutrons. A typical thermal column consists of a block of graphite of dimensions 8 ft × 5 ft × 2 ft containing a number of holes in which the various detectors are placed. Figure 3.1 shows such an arrangement.

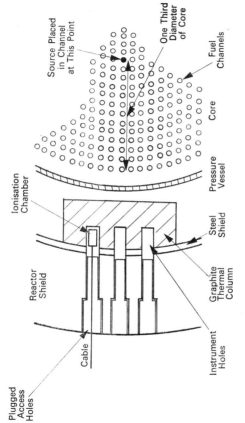

FIG. 3.1. Positioning of instruments and neutron source in a large natural uranium power reactor

3.4. USE OF A NEUTRON SOURCE

It was pointed out earlier that complete flux coverage was of vital importance in the start-up of reactors. Therefore there must be sufficient thermal neutrons at the detectors to give an adequate flux measurement with the reactor completely shut down. When reactors are in the completely shut-down condition the neutron flux level in the core will be low. For reactors fuelled with natural uranium a central neutron flux of the order of 20 n/cm^2 sec can be expected due to the spontaneous fission of uranium-238. If instruments are placed at the edge of the core, where a flux level of the order of 2 n/cm^2 sec could be expected, then an adequate measurement could be made. However, power reactor instruments are usually placed some distance from the core in an external thermal column. The flux level in the thermal column will of course be much reduced. The magnitude of this reduction is specified as a thermal column ratio for a particular reactor. This is defined as the ratio of the neutron flux at the reactor centre to the flux in the thermal column. Individual values of this ratio vary from reactor to reactor, but are usually of the order of 2000 to 5000. Thus if the central flux is 20 n/cm^2 sec the flux at the detectors for a thermal column ratio of 2000 will be 10^{-2} n/cm^2 sec which is too low for adequate measurement.

The spontaneous fission rate for uranium-235 is approximately a factor of 10^4 less than uranium-238 and therefore the flux at the centre of a reactor fuelled with highly enriched uranium is very small. Thus, although these reactors are smaller in physical size enabling instruments to be positioned near to the core, the flux level is still too low for measurement.

Hence the flux level in reactor cores must be increased in some way to enable sensible readings to be obtained at complete shut-down.

Another important aspect of flux measurement at shu-t down is the effect of gamma rays on the measuring instruments. Many instruments are sensitive to gamma rays and ionisation currents are produced in intense fields of gamma radiation. Initially the gamma rays produced in a reactor core are not of a sufficient intensity to effect the measurement of the neutron flux. However, as operation proceeds the gamma background from the core at shut-down increases as the number of gamma-active fission products increase. The relative neutron and gamma sensitivities for a typical ionisation chamber are such that the ratio of neutron sensitivity to gamma sensitivity is of the order of 10^3. As we have stated at shut-down the neutron flux level is very low, whereas after long periods of operation the gamma background from fission products and activated core components (i.e. steel pressure vessel) may be sufficiently high to give a reading on the measuring instruments. Hence on control rod withdrawal the instrument reading does not alter until the neutron flux level is high enough to exceed the measured gamma level, giving rise to a blind start-up.

In order to overcome these problems it is usual to load a neutron source into the core.[5] The neutrons produced by this source are multiplied by the fuel in the core and so the level of the neutron flux is raised to provide an adequate flux at the detectors.

3.4.1. Positioning of a Neutron Source

The position of the source in the reactor with relation to the neutron detectors is extremely important. The source must be positioned so that the neutrons emitted pass through some of the nuclear fuel before reaching the detector. If the detector "sees" the source neutrons directly, then no change in measured flux would occur until the core neutrons had reached a higher level and again a blind start-up would result. If the source is placed too far from the detectors then the flux level will again be too small for adequate measurement. A compromise must be adopted.

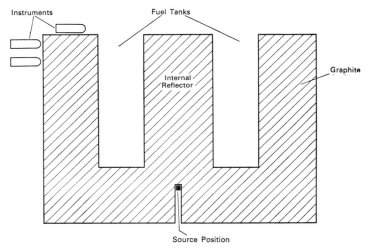

FIG. 3.2. Positioning of instruments and neutron source in a small research reactor

A typical arrangement for a large natural uranium reactor is shown in Fig. 3.1. By experience it has been found that the best position for the active source in this case is at a distance of one-third of a diameter from the edge of the core nearest to the thermal column. Figure 3.2 shows a typical arrangement for a small research reactor fuelled with enriched uranium.

3.4.2. Types of Source

The type and strength of neutron source varies according to the characteristics of the reactor. The source must, of course, be of a sufficient strength to raise the flux to the required level. For small research reactors a source emitting of the order of 10^5 to 10^6 n/sec is adequate whereas for large power reactors sources emitting 10^8 to 10^9 n/sec may be required. In addition it is often a requirement in a research reactor to be able to unload the source and hence ease of handling is of prime importance.

In view of the problems associated with the build-up of a large gamma background in a power reactor a very intense neutron source would be required. It is not feasible to manufacture a source of high intensity and in any case handling problems would be complex. Consequently it is usual to use either an inactive source or a source of comparatively low activity which will be activated by the operation of the reactor itself. Hence, as the gamma background builds up, the source activity also increases to maintain adequate neutron measurements at shut-down. If the source used is activated by the reactor then the radioactive properties of the source materials must be such that it can be activated in a reasonable time and also such that the induced activity will not decay too quickly in the event of a long period of reactor shut-down.

All sources used in reactors rely on the (α, n) or (γ, n) reactions with beryllium and are manufactured in the form of a central cylindrical α or γ emitter surrounded by a tube of beryllium or as a mixture of the α or γ emitter and beryllium. The complete source is often enclosed in a stainless steel sheath.

Sources for power reactors

Two possible sources may be used these being antimony–beryllium (half-life 60 days) and sodium–beryllium (half-life 15 hr). On account of the short half-life of the sodium–beryllium source it is more usual to use antimony–beryllium.

Sources for research reactors

The chief requirements in this case are long half-life combined with a low gamma ray emission to ease handling problems. The following sources may be used:

    (a) Plutonium–beryllium   — half-life 24,000 years
                                          — negligible gamma emission
    (b) Polonium–beryllium   — half-life 138 days
                                            — negligible gamma emission
    (c) Radium–beryllium   — half-life 1620 years
                                            — high gamma emission
    (d) Americium-beryllium — half-life 458 years
                                            — low gamma emission

The most frequently used sources are (a), (b) and (d) on account of the low gamma emission. The long half-life of plutonium–beryllium and americium–beryllium make them particularly useful.

3.5. USE OF LEAD-SHIELDED IONISATION CHAMBERS

It is common practice to surround ionisation chambers with a layer of lead or line the thermal column holes with lead in order to reduce the gamma ray field at the instrument. If 4 in. of lead is used, then the gamma flux at the ionisation chamber will be reduced by a factor of about 100, whereas the neutron flux will only be reduced by about 0·3.

3.6. TYPES OF INSTRUMENT FOR NEUTRON MEASUREMENT

Two types of instrument are commonly used for the measurement of neutrons.[6, 7, 8] These are:

    (a) Mean current ionisation chambers and
    (b) Pulse counters.

Neutrons are detected in both instruments by the ionisation produced as a result of the interaction of thermal neutrons with boron producing an α-particle in accordance with the reaction

$$_5B^{10} + _0n^1 \rightarrow _3Li^7 + _2He^4$$

or as a result of the fission of uranium-235 with the production of highly ionising fission products.

3.6.1. Ionisation Chambers

A typical ionisation chamber is shown in Fig. 3.3. It consists of two concentric cylidrical aluminium electrodes contained in an aluminium gas-tight container. The inner surface of the outer electrode and the outer surface of the inner electrode

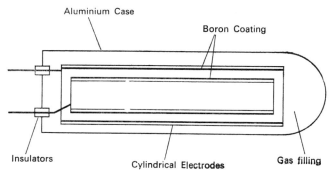

Fig. 3.3. Typical ionisation chamber

are coated with boron and the container is filled with hydrogen or argon. An alternative scheme utilises boron trifluoride gas in the container with uncoated electrodes. The neutron sensitivity of the ionisation chamber is dependent upon the area of the boron coating, the degree of enrichment of the boron with the isotope boron-10 (18·8% natural abundance) and the gas pressure within the container.

The sensitivity to gamma rays may be reduced by the use of a suitable gas in the container—hydrogen being the best choice for this purpose. Table 3.1 summarises the various properties of a number of ionisation chambers.

TABLE 3.1. TYPICAL CHARACTERISTICS OF IONISATION CHAMBERS

Type	Area of coating (cm²)	Sensitive volume for BF_3 gas (cm³)	Gas pressure hydrogen or BF_3 (cm Hg)	Neutron sensitivity (amps per unit flux)	Gamma sensitivity (amps per r/hr)	Gamma/ neutron sensitivity ratio
RC1	500	—	150	6×10^{-15}	6×10^{-12}	10^3
RC1	500	—	15	$1\cdot5\times10^{-15}$	6×10^{-13}	400
RC2	200	—	150	$2\cdot4\times10^{-15}$	$3\cdot3\times10^{-12}$	10^3
RC2	200	—	15	$0\cdot6\times10^{-15}$	$3\cdot3\times10^{-13}$	550
RC1	—	180	30	$2\cdot9\times10^{-14}$	$1\cdot35\times10^{-11}$	460
RC2	—	100	30	$1\cdot6\times10^{-14}$	$0\cdot75\times10^{-11}$	460
—	640 (96% B10)	—	225	$2\cdot6\times10^{-14}$	$6\cdot5\times10^{-12}$	250
—	640 (96% B10)	—	75	$1\cdot7\times10^{-14}$	$2\cdot2\times10^{-12}$	130
—	640 (96% B10)	—	25	$5\cdot7\times10^{-15}$	$7\cdot3\times10^{-13}$	128

3.6.2. Gamma Ray Compensation

The difficulties associated with the measurement of a neutron flux in the presence of a large gamma background at shut-down have already been discussed. It is important to attempt to reduce the gamma sensitivity of the low-power measuring instruments as much as possible. As can be seen from Table 3.1 the gamma sensitivity is reduced in ionisation chambers using coated electrodes, enriched boron and with reduced gas pressures.

The effect of gamma rays may be reduced considerably by a technique known as gamma ray compensation.[9] This is carried out by using two ionisation chambers—one measuring the effects of neutrons and gammas and the other measuring the gamma effect only. By arranging for the ionisation currents produced from the two chambers to oppose one another then the resultant current will only be due to the neutron flux. A typical gamma compensated ionisation chamber is shown

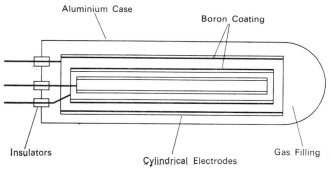

Fig. 3.4. Typical gamma compensated ionisation chamber

in Fig. 3.4. This arrangement consists of three concentric cylindrical electrodes. The inner surface of the outer electrode and the outer surface of the middle electrode are coated with boron. Therefore the current between electrodes 1 and 2 will be due to neutrons and gammas and between 2 and 3 due to gammas only. Compensation is never perfect due to the fact that there will be a gamma flux gradient across the ionisation chamber. In any case it is usual to arrange for the chambers to be undercompensated by about 3% as it is essential to prevent the gamma current produced being greater than the neutron current.

3.6.3. **Fission Type Ionisation Chambers**

These instruments are similar in design to the previous type being coated with a layer of uranium-235 instead of boron. The case material is usually stainless steel and the gas-filling argon. These instruments are not used so frequently as the boron type as they suffer from the disadvantage of a background count from the natural α-activity of the uranium and also from the build up of fission product activity following irradiation.

3.6.4. **Pulse Counters**

The most commonly used pulse counters in reactor instrumentation are the boron trifluoride proportional counter and the fission counter. They are used for the measurement of small fluxes in the range where the currents produced in the mean current ionisation chambers are too small for measurement. The principle of the proportional counter is to produce pulses as a result of irradiation and to use pulse amplification techniques to obtain a measurable signal. A typical BF_3 counter is shown in Fig. 3.5. It consists of a

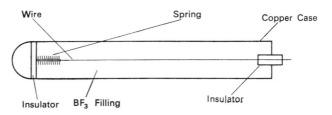

FIG. 3.5. Typical BF_3 counter

copper tube forming one electrode and a fine insulated wire at the centre of the tube, forming the other electrode, held at high voltage. The tube is filled with BF_3 gas. The central wire s at a high voltage to ensure that secondary ionisation occurs and a large pulse is obtained. A typical counter will be 7 in. long and 1 in. diameter and will be filled with enriched boron trifluoride at 40 cm of mercury pressure. A neutron sensitivity of the order of 3 counts/sec per unit flux will be achieved up to a maximum counting rate of the order of 10^4 to 10^5 counts/sec.

The pulse height obtained due to neutrons is found to be approximately 100 times that due to gamma rays and so by suitable pulse height discrimination the effects of gamma rays may be eliminated completely.

Fission counters are constructed in a similar way to BF_3 counters except that the counter is coated internally with uranium-235 and argon is used as a gas filling. The background due to the natural α-activity of uranium can be overcome by pulse height discrimination. Sensitivities of the order of 0·1 counts/sec per unit flux may be achieved with fission counters. These instruments are extremely insensitive to gamma rays and have the advantage that they will operate in gamma fields up to 10^5 r/hr. This coupled with the fact that they can be made to operate in temperatures up to 900°C makes them highly suitable as in-core instruments.

3.7. INSTRUMENTS ASSOCIATED WITH IONISATION CHAMBERS AND PULSE COUNTERS

As we have seen ionisation chambers are used to obtain a D.C. current and pulse counters a pulse as a result of a field of radiation. Ionisation chambers are not suitable for use in very low fields of radiation on account of the difficulty of measuring the small resulting D.C. current. In the case of small radiation fluxes pulse counters using pulse amplification techniques are more suitable.

The instrumentation used in conjunction with the two types of instrument will be discussed briefly.

3.7.1. Linear Amplifiers

D.C. linear amplifiers are used in conjunction with ionisation chambers in order to amplify the signal obtained sufficiently to operate indicating and recording instruments. Linear channels usually provide some sort of moving coil instrument to give a direct indication of the ionisation current and also operate a recorder producing a continuous record of the measured currents.

3.7.2. Backed-off Linear Amplifiers

A reactor operating at power will be driven by observation of the variation of a linear power recorder or indicator. It is therefore useful and convenient to provide an instrument which measures the linear variation in power from a pre-determined balance position. For this purpose a linear amplifier is used in conjunction with a variable stabilised current or voltage supply to back off the input current from the ionisation chamber. Hence at the balance power such an instrument would read zero and any variations from this position will be indicated. Instruments usually cover ranges between $+2$ to 5% to zero to -2 to 5% of full power.

3.7.3. Logarithmic Amplifiers

It is useful during reactor operation to provide a record of the complete power variation on a single instrument in order to have a convenient and easily obtainable record of the total irradiation. In order to achieve this it is necessary to cover a range of 7 or 8 decades. This can be carried out by the use of a logarithmic amplifier in which the output signal feeding an indicator or recorder is proportional to the logarithm of the input current from the ionisation chamber.[10]

Figure 3.6 shows typical characteristics of a diode. It is found that in the retarding field region (i.e. negative anode voltage) it is possible to obtain a current which varies exponen-

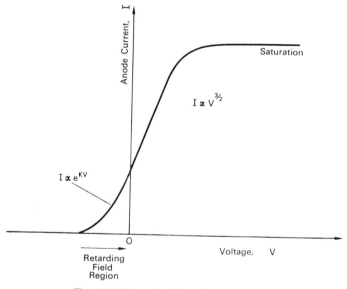

FIG. 3.6. Characteristics of a diode valve

tially with the voltage. A current can be obtained with negative voltages up to approximately 1·2 V and on account of the exponential relationship between current and voltage in this region it is found that a voltage change of approximately 0·2 V produces a current change of one decade. Thus by using a diode in place of a measuring resistor in a D.C. linear amplifier system, operation over 6 or 7 decades may be achieved in a region where the output is proportional to the logarithm of the input.

The equation relating the current and voltage in the retarding field region of a diode is given as

$$I_v = I_0 \exp \frac{eV}{kT} \tag{3.1}$$

where I_v is the current at voltage V,

$\quad I_0$ is the current for $V = 0$,

$\quad e$ is the electronic charge,

$\quad k$ is Boltzmann's constant,

$\quad T$ is the cathode temperature.

This equation may be expressed as:

$$V = \frac{kT}{e} \log_e I_v - \frac{kT}{e} \log_e I_0$$

and

$$V \propto \log_e I_v \tag{3.2}$$

It can be seen that the proportionality constant in the above equation contains T, the cathode temperature, and that I_0 is a varying function of the cathode temperature. Consequently in order that equation (3.2) shall apply the cathode temperature must be kept constant by some form of heater stabilisation. In practice it is more usual to use a pentode operating under conditions of constant anode current together with heater stabilisation. In this case the input is fed to the grid of the pentode and the output obtained from the screen grid; the screen-cathode region acting as a diode.

3.7.4. Period Meters

The reactor period or reactor doubling time is an important parameter in reactor operation as it gives a direct indication of the change in reactivity produced due to a physical change in the reactor. All reactor systems provide suitable instruments for measuring and indicating the reactor period or doubling time.[11] Logarithmic amplifiers may be used to provide this indication and are then called period meters.

Above criticality the reactor flux rises exponentially in accordance with the equation

$$\phi_t = \phi_0 \exp \frac{t}{T} \qquad (3.3)$$

where ϕ_t and ϕ_0 are fluxes at times t and $t = 0$, and T is the reactor period.

In terms of a doubling time T_D; then $T_D = T \log_e 2 = 0.693T$. If the exponentially rising flux is measured with an ionisation chamber and the resulting current fed to a logarithmic amplifier then the output voltage V will be given by

$$V \propto \log_e I_t; \quad \text{assuming that} \quad I_t \propto \phi_t.$$

Then
$$V = K \log_e \left(I_0 \exp \frac{t}{T} \right)$$

or
$$V = K \log_e I_0 + K \frac{t}{T}$$

where K is the amplifier constant.

By differentiating with respect to t, we get

$$\frac{dV}{dt} = \frac{K}{T}$$

i.e.
$$\frac{dV}{dt} \propto \frac{1}{T} \quad \text{or} \quad \frac{1}{T_D} \qquad (3.4)$$

Hence by differentiating the output signal from the logarithmic amplifier using an R.C. circuit a signal inversely proportional to the reactor period or doubling time is obtained which may be used to provide indication. Period meters usually provide indication over a range of -30 to ∞ to $+3$ sec. The signals are also used to provide doubling time warnings and trips.

3.7.5. Pulse Amplifiers

Pulse amplifiers are used with BF_3 counters and fission counters in order to amplify the pulse produced to operate indicators and recorders. Pulses obtained from counters are usually of the order of 0·1 mV. Pulse amplifiers are used in conjunction with pulse amplitude discriminators which require signals of the order of 10 to 30 V. Hence pulse amplification of the order of 10^5 is provided with the amplifiers.

Logarithmic ratemeters are also used in conjunction with the pulse counters. These instruments operate on similar principles to logarithmic amplifiers and produce a signal which is proportional to the logarithm of the input. They are used to operate indicators and recorders. They may also be used in conjunction with a differentiating circuit to provide period indication and protection in a similar manner to logarithmic amplifiers.

3.8. TYPICAL ARRANGEMENTS FOR REACTOR POWER INSTRUMENTATION

In order to provide adequate measurement of reactor power and neutron flux from complete shut down to full power and for operation at full power several instruments are used having a variety of uses and covering different ranges.[12, 13, 14]

The following instruments and associated equipment are used.

3.8.1. Start-up Channels

Two BF_3 counters are used and placed in a hole in the thermal column. Provision is made to retract these counters at high powers. The signals from the counters are amplified by a head amplifier, which supplies the main pulse amplifier. The pulse amplifier provides signals to operate a logarithmic count rate meter and indicator. Period and excess count trips are provided.

3.8.2. Shut-down Channels

Three uncompensated ionisation chambers are placed in three thermal column holes. They each feed signals to a linear amplifier providing trips at pre-set power levels.

3.8.3. Low Logarithmic Power Channel

One compensated ionisation chamber is used and placed in a lead-lined thermal column hole. A logarithmic amplifier, period meter, recorder and indicator are associated with this channel. The instruments provide power trips and period trips. Provision is made to retract the ionisation chamber at high powers.

3.8.4. Main Logarithmic Channel

One compensated ionisation chamber is used and placed in a lead-lined thermal column hole. A logarithmic amplifier, period meter, recorder, indicator, power and period trips are provided.

3.8.5. Linear Channel

One compensated ionisation chamber is used and placed in a hole in the thermal column. The signals are fed to a linear amplifier which operates a linear recorder.

3.8.6. Backed-off Channel

One compensated ionisation chamber is used and placed in a hole in the thermal column. Signals are fed to a linear amplifier used in conjunction with a stabilised voltage supply and a backed-off recorder.

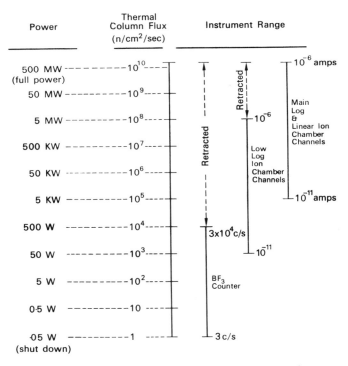

FIG. 3.7. Flux and power ranges covered by the nuclear instruments

The operating ranges covered by these instruments with relation to the reactor power and thermal column flux are indicated in Fig. 3.7.

Let us assume that the maximum power of the reactor is 500 MW corresponding to a central flux of 2×10^{13} n/cm^2 sec. At shut-down the power and flux are 0·05 W and 2×10^3 n/cm^2 sec respectively with the installed neutron source fully activated. If the thermal column ratio is 2000 then the neutron flux in the thermal column at full power and shut down will be 10^{10} and 1 n/cm^2 sec respectively. Ionisation chambers with a neutron sensitivity of 2×10^{-14} amps per unit flux and a gamma sensitivity of 2×10^{-12} amps/r hr and BF$_3$ counters with a neutron sensitivity of 3 c.p.s. per unit flux are placed in the thermal column. The instruments will cover the following ranges.

(a) *At shut-down*

BF$_3$ reading	3 c.p.s.
Ionisation chamber current	2×10^{-14} A (neutron)
	2×10^{-13} A (gamma)
	(lead-lined hole)

(assuming a gamma shut-down flux of 10 r/hr or 0·1 r/hr under 4 in. of lead).

At this level all measurements are made using the BF$_3$ counters only.

(b) *At 50 W power*

BF$_3$ reading	3×10^3 c.p.s.
Ionisation chamber current	2×10^{-11} A (neutron)
	(gamma swamped)

Both instruments used at this power level.

(c) *At 500 W power*

BF$_3$ reading	3×10^4 c.p.s.
(maximum	5×10^4 c.p.s.)
Ionisation chamber current	2×10^{-10} A

BF_3 counters withdrawn at this stage and measurements made with low logarithmic ionisation chambers.

(d) *At 5 kW power*

Low logarithmic ionisation chamber current 2×10^{-9} A

Main logarithmic and linear ionisation chamber currents
2×10^{-11} A

All above instruments used at this power level.

(e) *At 5 MW power*

Low logarithmic ionisation chamber current 2×10^{-6} A
(maximum)

Main logarithmic and linear ionisation chamber currents
2×10^{-8} A.

Low logarithmic ionisation chambers withdrawn at this level and the remaining power range up to 500 MW is measured using the main logarithmic and linear channels. The ranges covered are illustrated in Fig. 3.7.

REFERENCES

1. GLASSTONE, S. and SESONSKE, A. *Nuclear Reactor Engineering.* Van Nostrand (1963).
2. JERVIS, M. W. *Nuclear Reactor Instrumentation.* Temple Press (1961).
3. SCHULTZ, M. A. *Control of Nuclear Reactors and Power Plants.* McGraw-Hill (1955).
4. COX, R. J. and SANDIFORD, K. R. Reactor control and instrumentation. *J. Brit. Nuc. Eng. Conf.* **2**, 2 (1957).
5. Role of neutron source in reactor safety. *Nuclear Safety* **4**, 3 (1963).
6. ALLEN, W. D. *Neutron Detection.* Newnes (1960).
7. ABSON, W. *et al.* Neutron flux instrumentation for research and power reactors. Geneva Conf. Paper 15 P/56 (1958).
8. ABSON, W. and WADE, F. Nuclear reactor control ionisation chambers. *J. Brit. Nuc. Eng. Conf.*, July (1956).
9. GRAY, A. L. Gamma compensated ionisation chambers for reactor control. *Nuclear Power*, Dec. (1959).

10. BARROW, B. B. The log-diode counting ratemeter and period meter. *Advances in Nuc. Eng.* II. Pergamon (1957).
11. Role of log N period meter in reactor protection. *Nuclear Safety* **4,** 3 (1963).
12. SCHULTZ, M. A. Reactor safety instrumentation. *Nuclear Safety* **4,** 2 (1962).
13. SIDDALL, E. Safe and reliable reactor protection. *Nucleonics* **15,** 6 (1957).
14. JACOBS, I. M. *Safety Systems for Nuclear Power Reactors.* A.I.E.E. (1957).

CHAPTER 4

Reactor Start-up

4.1. REACTOR START-UP SEQUENCE

The start-up sequence for any particular reactor is dependent on the type and characteristics of the reactor, the coolant circuits, the heat exchange system and the power generation system. A sequence is determined by considering the controls available and their effect on reactor conditions, the order of carrying out start-up operations to provide the maximum efficiency with maximum safety, the effect of rise of temperature on reactor and plant conditions and the effect of the steam turbines on reactor performance. Obviously in a text of this size it is impossible to deal with every type of system, but a number of general methods will be discussed and a fairly detailed start-up procedure will be given for a graphite-moderated, gas-cooled, natural uranium power reactor.

In most research reactors the main control from the point of view of start-up is provided by a system of control rods or absorbers. A large number of preliminary checks have to be carried out especially with a facility which has a variable core structure and variable experimental facilities to ensure that conditions are ready for reactor start-up. It is usual to provide a series of interlocks which prevent movement of the control system until certain items of the reactor system and associated plant are in the correct operational condition. These interlocks are necessary in order to prevent a dangerous situation arising as a result of mal-operation. Hence for research reactors the start-up sequence would commence with tests necessary to ensure that correct operating conditions had been established before commencing any movement of the control system. The reactor would then be started up in accordance with a pre-determined pattern by withdrawing the control system in carefully controlled stages. At each stage observations would be carried out to verify that normal reactor conditions had been achieved at each stage. Temperatures would be allowed to rise in accordance with the plant limitations and eventually the power would rise to its normal level.

In the case of power reactors, reactor conditions and hence the start-up sequence are dependent upon four factors. These are:

(a) The control system.
(b) The coolant flow.
(c) Variations in steam conditions in the heat exchangers and steam turbines.
(d) Plant limitations.

The effect on the plant and the relative importance during reactor start-up of these factors will be considered in turn.

4.1.1. The Control System

All reactors have a control system which consists of a number of absorbers, usually in the form of rods or plates, which are withdrawn and inserted to vary the core reactivity and hence change the operating conditions.[1, 2, 3, 4] By the gradual withdrawal of the control system the reactor may be brought to criticality and eventually to full power.

4.1.2. Coolant Flow

Any change in the coolant flow through the core may have a considerable effect on the reactivity of the reactor and so may be used to exert controlling action during start-up. In the case of reactors with negative temperature coefficients of reactivity a large variation in reactor power may be achieved by changing the coolant flow rate through the core. For instance, if the reactor is initially balanced at low power and low coolant flow then the power can be raised merely by raising the rate of flow of coolant without need to withdraw the control system at all. Power rises with coolant flow due to the negative temperature coefficient of reactivity. Increasing the coolant flow tends to reduce reactor temperatures, which leads to a positive change of reactivity with a corresponding increase in neutron flux and hence a larger heat output from the reactor. Control of the reactor power during start-up by coolant flow variation is used in gas-cooled reactors to enable power to be raised without causing large rates of change of reactor temperatures. For this reason the gas circulators are provided with a system for varying their speed. A variety of methods have been employed, the most commonly used being the Ward–Leonard system, the centrifugal clutch method and the steam turbine drive system.[5, 6] Gas circulator speed variation is required to vary the coolant flow continuously over the range from full flow to 10% flow.

4.1.3. Variations in Steam Conditions

Changes in steam flow rates and temperatures in both the heat exchanges and the turbines effect reactor conditions and consequently may be used to some extent as control parameters.[7] During start-up changes in reactor temperatures produce reactivity and hence power changes in the reactor. Changes in the steam conditions will produce temperature changes in the coolant outlet temperature from the heat exchanges and hence changes in the coolant inlet temperature to the reactor core. During reactor start-up changes in core temperatures are carefully controlled to prevent excessive thermal stressing of the core components. In order to obtain reliable information from temperature survey measurements and to exercise some degree of control over reactor temperatures it is essential to maintain a uniform coolant inlet temperature to the reactor core. This may be achieved by control of steam conditions. For instance, a flow control valve, controlled by the reactor operator, may be placed in the low-pressure steam line between the heat exchangers and the steam turbines. By varying the setting of this valve the coolant outlet temperature from the heat exchangers and hence the reactor inlet temperature may be varied and controlled to suit reactor operating conditions. Alternatively by combining the steam lines from the heat exchanges to several steam turbines and the use of dump condensers and variable steam flow valves similar control may be exerted on reactor inlet temperatures.

4.1.4. Plant Limitations

Reactor start-up and operation are governed by the limitations and restrictions which may be placed on certain items of the plant. In general limitations will usually apply to fuel element temperatures and rates of change of temperature, temperature differentials in certain parts of the plant, and coolant and steam pressures.

Limitations to these items are necessary from the point of view of the overall safety of the reactor and its components. The restrictions ensure that there is an adequate safety margin during operation so that as a result of the maximum credible accident there can be no possibility of a core melt-out or large release of activity into the surrounding area.

Operating limits are usually placed on the following:

1. Maximum fuel can temperature.
2. Rate of change of fuel can temperature.

3. Maximum indicated moderator temperature.
4. Maximum coolant pressure.
5. Maximum reactor vessel temperature.
6. Rate of change of reactor vessel temperature.
7. Rate of release of reactivity by movement of the control rods.
8. Reactor vessel–coolant circuit duct differential temperatures.

4.2. TYPICAL START-UP SEQUENCE FOR A GAS-COOLED REACTOR

A typical start-up sequence using the above methods of control would be:

1. Raise the reactor core and component temperatures to the normal coolant inlet temperature.

This would be achieved by operating the gas circulators at their minimum speed and allowing the coolant to flow through the core. Temperatures will rise due to the heat input from the gas circulators (usually of the order of 1 to 2 MW per circulator). Temperatures may easily be raised to about 140°C by this method without the need for any additional heat supply.

2. Withdraw the control rods in stages to achieve criticality. Maintain reactor temperatures at the normal inlet temperature by gas circulator operation at minimum speed.

3. Continue the control rod withdrawal and allow the reactor flux to rise. Eventually nuclear heat will be produced. The control rod withdrawal is continued so that fuel element temperatures increase in accordance with the plant limitations.

4. Allow the fuel element temperatures to rise to the maximum value and balance by movement of control rods. During this process the gas circulators are still maintained at mini-

mum speed and all steam produced is usually dumped by use of a dump condenser. The gas inlet temperature is now maintained at a uniform value by variation of the steam flow to the dump condenser.

5. Raise reactor power to maximum value in accordance with the plant limitations by raising the gas circulator speed from minimum to maximum. During the power rise fuel temperatures are maintained by slight adjustments of the control rods. The reactor inlet temperature is maintained by variation of the low pressure steam flow.

During prolonged operation at power movement of the control rods is necessary to maintain reactor conditions at full power as the following changes take place:

(a) Graphite temperatures take approximately 5 hr to reach equilibrium.
(b) Xenon poisoning builds up to its equilibrium value in 3 days.
(c) Samarium poisoning effects occur over a period of approximately 14 days.
(d) Long-term changes and build-up of plutonium.

4.3. THE IMPORTANCE OF A DUMP CONDENSER

Any change in steam conditions at the steam turbines in a power reactor will produce changes in heat exchanger temperatures which lead to changes in reactor temperatures. Consequently reactor temperatures can be effected by steam turbine conditions. On base load stations frequency changes in the National Grid System will cause the turbine governor valves to alter the steam supply to the turbines in order to maintain a constant frequency. Changes of this nature in the steam supply to the turbines are occurring continuously and therefore reactor conditions may fluctuate. This will cause thermal cycling of the fuel element temperatures and will produce very difficult operating conditions. It is therefore important to avoid interference to reactor temperatures by the steam generation system. This is achieved at nuclear power plants by the installation of a fairly large steam dump facility. It is usual to install a sufficient number of dump condensers to achieve a dump capacity of at least 10 to 12% of the total heat output of the station. The dump condensers are then used in association with flow control valves and the steam turbines to take up steam supply variations to the turbines without changing steam conditions at the heat exchangers. Hence by the use of a dump facility it is possible to avoid continual changes in reactor conditions as a result of steam turbine variations.

Another extremely important need for a dump facility at a nuclear power plant occurs during start-up of the plant. Steam generation in the heat exchangers occurs gradually and slowly due to the slow rate of increase of reactor temperatures. Consequently initially very poor quality steam is generated which is not suitable for the steam turbines and may in fact cause damage to the turbines. Hence steam is supplied to the dump condensers during start-up until acceptable steam conditions are achieved for the turbines.

4.4. OVERALL STATION CONTROL[8]

A large nuclear power station may well consist of two reactors, each having 4, 6 or 8 heat exchangers supplying steam to a turbine house with 4 or 6 turbines (2 or 3 associated with each reactor). It is found advantageous from the point of view of overall station control to have a single central control room with sufficient instrumentation to enable the whole station (reactors, heat exchangers and steam turbines) to be controlled.

In addition it is usual from the point of view of reactor temperature stability to combine the steam lines from all the heat exchangers, steam turbines and dump condensers. A typical arrangement for a two-reactor—16 heat exchanger—four-turbine station is shown schematically in Fig. 4.1.

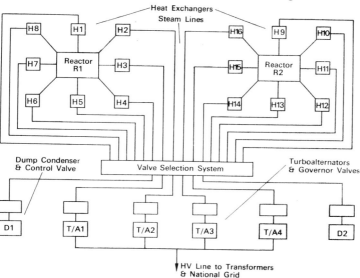

FIG. 4.1. Schematic diagram illustrating typical link-up between reactors, heat exchangers and turbo-alternators

By combining the steam lines and using the dump condensers and a system of variable steam valves it is often feasible to accommodate the complete trip out of one turbine without having to shut down a reactor. Variations in steam flow to the turbines as a result of grid frequency changes may easily be accommodated with this system without effecting temperature conditions in the reactors.

4.5. OVERALL CONTROL OF DIFFERENT TYPES OF REACTOR

The reactor control system must bring the reactor and its auxiliaries from shut down to full power making due allowance for conditions in all plant components and the demands imposed by the external load. Consequently as we have mentioned in previous sections the effect of the heat exchanger/

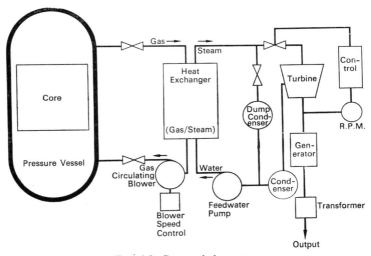

FIG. 4.2. Gas-cooled reactor

steam turbine part of the plant and their requirements and limitations must be included in any proposed start-up sequence. The control system should be designed to limit the disturbances produced in the reactor by varying conditions in the power plant. The effect on the reactor of loss of load due to turbine or transmission line faults should be estimated.

The specific parameters which determine the overall control of a reactor and its plant, depend upon the reactor type and

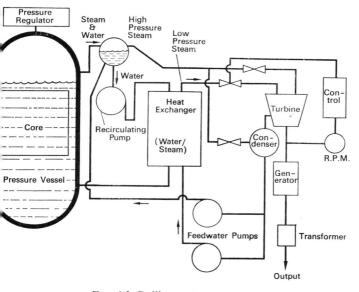

Fig. 4.3. Boiling water reactor

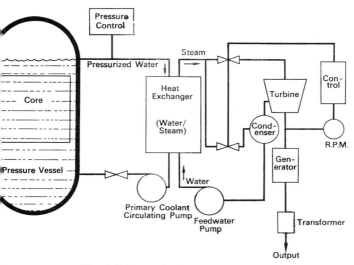

Fig. 4.4. Pressurised water reactor

its characteristics. A number of important control parameters
for four different types will be considered. These are the gas-
cooled reactor (Fig. 4.2.—already discussed in some detail

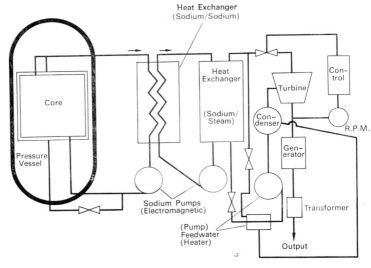

FIG. 4.5. Liquid–metal-cooled reactor

in other sections of this chapter), the boiling water reactor
(Fig. 4.3), the pressurised water reactor (Fig. 4.4) and the
liquid–metal-cooled fast reactor (Fig. 4.5).

(a) *Neutron flux*

The measurement of neutron flux is used on all four types
to determine the operating power level of the reactor and to
initiate excess power and period trips (see Chapter 3). This
method provides quick response to reactor disturbances and
is the only method that permits auto control and start-up from
the completely shut-down condition. Continual calibration

is required at high-power operation in terms of a reactor heat balance (see Chapter 9) as the ratio of the measured flux using instruments placed at the edge of or outside the core to the mean core flux varies with reactor conditions.

(b) *Coolant outlet temperature*

In the case of the gas-cooled reactor and the pressurised water reactor the measurement of outlet temperature provides information about the mean core flux and can therefore be used to determine the average thermal conditions in the reactor. As mentioned previously, channel outlet temperature measurement is used extensively in gas-cooled reactors for flux shaping and for determining the total power generated in the reactor. In all types of reactor the outlet temperature is used as a backup for neutron flux measurements to sense power excursions and to determine the rate of reduction of coolant flow rate.

(c) *Coolant inlet temperature*

The coolant inlet temperature is used to determine the coolant temperature rise through the core (i.e. $T_{outlet} - T_{inlet}$). Obviously if coolant outlet temperature variations are used to determine flux changes in the core and for optimising power production then it is essential to maintain the inlet temperature at a constant value. Systems are designed so that heat exchanger and steam variations do not cause undue reactor inlet temperature changes. The gas-cooled reactor has been considered in detail. Similar methods are employed on other systems using dump or main steam condensers and variable flow valves to exert control (see Figs. 4.2, 4.3, 4.4 and 4.5).

(d) *Coolant flow rate*

The coolant flow rate in all reactors is a basic safety parameter. Loss of coolant could lead to core melt out and must be prevented at all costs. The flow rate is measured on water reactors by determination of the pressure drop across an orifice plate or from the coolant pump operating conditons. In water reactors the coolant flow is usually kept constant, whereas in gas-cooled reactors the flow is varied to control reactor power (as discussed previously). In boiling water reactors the coolant flow is often varied as a means of load following.

(e) *Coolant pressure*

In gas-cooled reactors the coolant pressure is used as a safety parameter to detect loss of coolant. Rapid rate of change of pressure is used to initiate reactor shut down. In the pressurised water reactor coolant pressure measurement is necessary to warn of excess pressure and possible system rupture. Also in this case rapid loss of pressure can reduce the saturation temperature below the existing temperature and coolant boiling can start. It is usual to design the control system to prevent boiling and an emergency cooling system is incorporated for this purpose.

(f) *Steam flow*

The amount of steam produced is often used to determine the reactor heat output. This is of particular importance in the case of boiling water reactors. Other methods are more accurate in other cases.

(g) *Steam pressure*

The steam pressure at the turbine is a control parameter in any thermal power plant. Pressure changes are used in nuclear installations to regulate the position of a valve or system of valves controlling the steam flow between the turbine and a dump condenser. Steam pressure is more important in boiling water reactors and is used as a control and a safety parameter. Safety implications arise in this case as an increase in steam pressure will cause void collapse and hence a power excursion (with a negative void reactivity coefficient). Steam pressure regulating devices must be used to avoid this possibility.

(h) *Feedwater flow*

Feedwater flow and associated reactor water level are particularly important in boiling water reactors (see Fig. 4.3). In this case the feedwater returns directly from the condenser to the reactor and hence feedwater temperature changes due to turbine load changes effect reactor coolant temperatures. In addition the head of water above the reactor core is effected by the steam void volume which is dependent upon turbine load conditions. Hence control of reactor core temperatures as the turbine load varies is difficult. Some degree of control is obtained by altering the feedwater flow to regulate the difference between steam flow out and water flow in.

i) *Turbine speed*

The speed of the turbine is a measure of the load requirement. The variation of turbine speed is used to vary a governor valve to control the steam flow to the turbine. Governor valve control is essential on base load stations.

4.6. PROCEDURE FOR THE START-UP OF A GAS-COOLED NUCLEAR POWER STATION

In order to illustrate the points already discussed with regard to a typical reactor start-up, it is intended to outline the procedures involved in starting-up a large gas-cooled graphite-moderated reactor.

It is assumed that the maximum reactor heat output is 550 MW with a maximum fuel can temperature of 440°C, a reactor inlet temperature of 160°C, and a maximum electrical output of 140 MWe. A dump capacity of 12 % (i.e. 66 MWh) is available. The control system consists of 100 uniformly distributed control rods divided into four uniformly distributed groups, these being:

Safety group—20 rods
Shut-down group—36 rods
Start-up group—36 rods
Operating group—8 rods

The reactor instrumentation consists of a retractable in-pile fission counter for measurements at complete shut down, a retractable BF_3 counter, a retractable low log channel, a linear channel and a main log channel.

The following sequence would be adopted.

4.5.1. Preparations

1. Check that all power supplies are available.
2. Charge the gas circuits with coolant to 85 psig (normal operational pressure 125 psig).
3. Check that all shield plugs and shield facilities are closed.
4. Check that the neutron source and in-pile fission counter are installed and that all instrumentation is functional.
5. Ensure that the fuel charge/discharge machinery is available and functional.
6. Open the upper gas duct valves (outlet ducts) and close the lower gas duct valves (inlet ducts).
7. Check that the gas circulators are operational.

8. The H.P. and L.P. drums on the heat exchangers should be filled with water to the operating level.
9. The turbine house and cooling water equipment should be available and functional.
10. All reactor safety lines should be set and healthy.
11. All control rods should be fully inserted.
12. All other equipment associated with the operation of the station at power should be functional and tested.

4.5.2. Initial Operations

1. Raise the 20 safety rods.
2. Start the gas circulators at minimum speed (i.e. 10% coolant flow—300 r.p.m.).
3. Open bottom gas duct valves (inlet ducts).
4. Raise gas circulator speed to 20% coolant flow (600 r.p.m.).
5. Allow temperatures to rise by heat input from the gas circulators until uniform temperatures of 140°C are achieved.
6. Open the appropriate heat exchanger valves when pressures of the order of 10 psig are reached.
7. Maintain reactor temperatures at 140°C.

4.5.3. Approach to Criticality

1. Measure neutron flux with in-pile fission counter and BF_3 counter (in thermal column).
2. Raise all control rods by one-tenth of their total travel and check for slack wires to ensure that all rods have lifted.
3. Measure neutron flux with in-pile fission counter and BF_3 counter.
4. Raise the 36 shut-down rods in stages and obtain neutron flux measurements at each stage.

5. When in-pile fission counter channel reaches 10,000 c.p.s., cross-check with BF_3 counter measurements and then withdraw in-pile fission counter from the core.

6. Continue rod withdrawal until criticality is attained. This is achieved by withdrawing the rods until the neutron flux continues to rise on stopping the rod movement.

7. Continue rod withdrawal until the neutron flux is rising with a doubling time of approximately 30 sec. Stop rod movement. (When the 36 shut-down rods are fully withdrawn continue with the 36 start-up rods.)

8. Allow flux to rise until power reaches approximately 50 W and insert control rods to balance at this power level.

9. At 50 W cross-check all instruments and carry out a reactor temperature survey.

4.5.4. Raising to Power (see Table 4.1 and Fig. 4.2)

Stage 1—Raising reactor temperatures to 160°C

1. Continue rod withdrawal and raise power to 5 kW or the power level at which the BF_3 channel indicates 10,000 c.p.s. Balance at this level.

2. Cross check with low log channel and then withdraw BF_3 counters from the thermal column.

3. Continue rod withdrawal, raise power to 5 MW and balance.

4. Carry out instrument check and obtain a complete temperature survey.

5. Retract low log power instruments at this stage and continue with main log channel and linear channel.

6. Raise rods and note that at this stage nuclear heating commences. Regulate rod withdrawal so that the fuel element can temperatures rise at less than 4°C/min.

7. Blow through heat exchanger drains and warm L.P. and H.P. steam lines.
8. Start heat exchanger feed pumps.
9. Commence steam dumping using dump condenser.
10. As reactor gas inlet temperature reaches the operational value of 160°C progressively bring in the dump condenser steam flow valve as required to maintain the reactor gas inlet temperature at 160°C. Balance at this level.

Stage 2—Raising to 60 MW reactor power

1. Continue to allow temperatures to rise by rod withdrawal until a maximum fuel element can temperature of 240°C is achieved.
2. Carry out complete survey of reactor temperatures.
3. Raise gas circulator speed in small increments maintaining fuel element temperatures constant by rod adjustment if necessary.
4. Balance at 60 MW.
5. Maintain gas inlet temperature at 160°C by regulating steam discharge to dump condenser.
6. Survey reactor conditions (see Table 4.1).

At this stage the limit of the dump capacity (66 MW) is being approached and therefore before raising power any higher it is necessary to bring into operation a steam turbine. However, with a reactor gas outlet temperature of 215°C acceptable steam conditions will not have been achieved. Consequently the next stage of power raising is concerned with increasing reactor temperatures to achieve a higher gas outlet temperature but without raising power above 60 MW. This is attained by rod withdrawal and reduction in gas circulator speed.

Stage 3—Obtaining acceptable steam conditions at 60 MW reactor power

1. Continue rod withdrawal and raise the maximum fuel element can temperatures to 320°C.
2. At the same time maintain the reactor power at 60 MW by reduction in gas circulator speed.
3. Allow HP steam pressure to reach 300 psig and maintain it at this value by regulating the passage of the HP steam to dump.
4. Maintain the gas inlet temperature at 160°C by regulating the L.P. steam pressure.
5. Balance at 60 MW with the maximum fuel element can temperatures at 320°C.
6. Carry out complete temperature survey.
7. Prepare steam turbine for operation (i.e. raise vacuum, carry out mechanical run and dry out).
8. Load steam to turbine at rate of 1 MWe/min. Transfer H.P. steam from dump to turbine. Load and synchronise to 5 MWe. Continue to dump L.P. steam.

Stage 4—Raising gas outlet temperature to 300°C

1. Raise maximum fuel element can temperatures to 380°C by rod withdrawal at constant gas circulator speed.
2. Progressively load the turbine and maintain H.P. boiler steam pressure constant.
3. Balance at power and carry out complete survey of conditions.

Stage 5—Raising fuel element temperature to 420°C

1. Raise maximum fuel element can temperatures to 420°C (20°C less than limit) by rod withdrawal and adjustment of gas circulator speed.
2. Balance at 120 MW.

3. Load the turbine progressively to maintain H.P. steam pressure constant.

4. As the power increases to 120 MW gradually transfer L.P. steam from dump to turbine.

5. Maintain maximum fuel element can temperature at 420°C.

6. Maintain gas inlet temperature at 160°C by regulating the ratio of L.P. to H.P. steam pressure by means of the steam flow control valves.

7. Discard dump condenser completely.

8. Survey reactor conditions.

Stage 6—Raising reactor power to 250 MW

1. Raise gas circulator speed and allow power to rise to 250 MW.

2. Maintain maximum fuel element can temperature at 420°C by rod adjustment if necessary.

3. Carry out complete survey of reactor conditions.

Stage 7—Raising to maximum power

1. Raise maximum fuel element can temperature to maximum value of 440°C at 1°C/min.

2. Raise power by raising gas circulators to maximum speed and maintain maximum fuel element can temperature constant by rod adjustment if necessary.

3. Carry out survey of reactor conditions on attaining full power.

Details of typical operating conditions at each of the above stages are given in Table 4.1.

TABLE 4.1. TYPICAL OPERATING CONDITIONS DURING START-UP OF A GAS-COOLED NUCLEAR POWER REACTOR

Stage	Reactor power	Max. F.E. can temperature	Gas inlet temperature	Gas outlet temperature	Gas circulator speed	Coolant flow	Dump condenser	Steam turbine
1	5 to 10 MW	160°C	160°C	160°C	600 rpm	20%	5 to 10 MWh 60 MWh	—
2	60 MW	240°C	160°C	215°C	1200 rpm	40%		—
3	60 MW	320°C	160°C	265°C	800 rpm	26·7%	L.P. steam	5 MWe
4	80 MW	380°C	160°C	300°C	800 rpm	26·7%	L.P. steam	10 MWe
5	120 MW	420°C	160°C	330°C	1000 rpm	33·3%	—	30 MWe
6	250 MW	420°C	160°C	315°C	2300 rpm	76·7%	—	65 MWe
7	550 MW	440°C	160°C	345°C	3000 rpm	100%	—	140 MWe

REFERENCES

1. GLASSTONE, S. and SESONSKE, A. *Nuclear Reactor Engineering*. Van Nostrand (1963).
2. Manual for the operation of research reactors. *I.A.E.A. Technical Report Series* No. 37 (1965).
3. SCHULTZ, M. A. *Control of Nuclear Reactors and Power Plants*. McGraw-Hill (1955).
4. SCHULTZ, M. A. *Proc. Conference on Nuclear Engineering*. Berkeley (1953).
5. Calder Hall atomic power station. *Nuclear Engineering* **1,** 7 (1956).
6. BOWDEN, A. T. and MARTIN, G. H. Design of important plant items. *J. Brit. Nuc. Eng. Conf.* April (1957).
7. WOOLTON, W. R. Steam-cycle analysis. *J. Brit. Nuc. Eng. Conf.* April (1957).
8. THOMPSON, T. J. and BECKERLEY, J. G. *The Technology of Nuclear Reactor Safety*, Chapter 6. M.I.T. Press (1964).

CHAPTER 5

Pre-Nuclear Commissioning

COMMISSIONING is a vital part of reactor operation. The initial operation of a new nuclear facility must be carried out in carefully planned stages and each component tested separately to ensure maximum reliability and safety.[1] The aim of a commissioning programme is to obtain accurate information about the reactor core and its components prior to carrying out any full-scale operation. In this way operators become familiar with the behaviour of the plant in gradual stages and obtain experimental data about all items of equipment which provides a sound basis for all future operations.

It is usual to divide commissioning programmes into two separate parts; these being:

(a) Pre-nuclear phase or conventional phase (no nuclear fuel loaded).
(b) Nuclear phase (commencing with the loading of fuel).

The tests carried out in the pre-nuclear phase are aimed at testing all items of equipment associated with the operation of the reactor and to show that the plant is behaving in accordance with a detailed design specification. The proposed commissioning programmes are always submitted to the appropriate licensing authority for approval prior to commencing the tests. The results of the pre-nuclear phase tests must be analysed and evaluated before the loading of nuclear fuel commences. The plant components must function correctly and if any test indicates that this is not so then the equipment must be rectified before continuing with the commissioning programme.

Commissioning a nuclear reactor is always a lengthy operation in view of the necessity to test each component. Even on small reactors two or three months is occupied in carrying out the conventional and nuclear commissioning tests. It can take 9 to 12 months or even longer to commission a full-scale nuclear power reactor.

5.1. MECHANICAL AND ELECTRICAL TESTS

The number of experiments to be carried out to test the mechanical and electrical equipment will obviously be dependent on the type and complexity of the reactor system. In general a commissioning programme consists of testing each particular reactor component.

Tests will usually include the following items.

5.1.1. Functional Tests

A number of functional tests are always carried out to ensure that there are no unforeseen difficulties and to enable the operators to familiarise themselves with the operational procedures. These tests will include checks on

(a) Beam tube and plug removal.
(b) Erecting and dismantling all experimental facilities associated with the reactor.
(c) Fuel store and fuel-handling facilities.
(d) Coolant supplies and charging the coolant circuits.
(e) Control rods—testing of limit switches, position indicators, withdrawal times and drop times.
(f) Flux scanning equipment.

5.1.2. **Coolant Circuit Tests**

Tests will be carried out to check:

(a) Coolant flow in each fuel channel and through the core (Section 5.2).

(b) Coolant purity.

(c) Leak testing of the reactor vessel and all coolant circuits.

(d) Calibration of the coolant flow-measuring equipment.

5.1.3. **Thermocouple Tests**

All reactor thermocouples and temperature-measuring devices will be checked and calibrated.

5.1.4. **Electrical Tests**

The following checks will be carried out:

(a) The nuclear instrumentation will be tested by electrical injection tests and by exposure to a source of neutrons.

(b) Functional tests will be carried out on the interlock system. Faults will be simulated to ensure that the correct indications and actions take place.

(c) All control and safety devices will be checked and faults simulated to ensure correct operation.

On the satisfactory completion of the conventional commissioning programme the reactor will be ready for the initial loading of nuclear fuel.

5.2. COOLANT FLOW TESTS

In the case of power reactors it is extremely important to obtain accurate measurements of the coolant flow in each individual fuel channel. It is essential to achieve a maximum heat output from the core of a power reactor. This will be obtained if the highest possible channel outlet temperatures are attained for a given set of reactor conditions. With this aim in mind some degree of flux flattening is usually employed by the use of fixed and removable absorbers placed in the central region of the core. By this means the region of high neutron flux and hence high fuel and coolant channel outlet temperatures is extended giving higher heat output.

Higher coolant channel outlet temperatures may also be achieved by controlling the coolant flow in each individual channel. For instance, if a system of variable orifices or restrictions are placed at the bottom of each channel, then these could be adjusted to alter the coolant flow to suit the neutron flux and so give maximum outlet temperatures. Assuming perfect flattening Fig. 5.1 indicates a typical radial flux distribution. Maximum coolant flow should be achieved in the flattened region (up to radius R_F). In the region from radius R_F

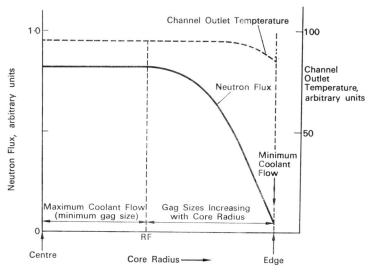

FIG. 5.1. Flux and channel outlet temperature flattening

to the edge of the core the coolant flow should gradually be reduced with minimum flow at the edge.

The ideal arrangement would be to install a system of variable gags or orifices at the bottom of each coolant channel which could be varied from outside the reactor. If the coolant outlet temperature from each channel was measured then the operators could adjust the gags to achieve maximum temperatures for any particular neutron flux distribution. In this ideal situation it would be possible to dispense with any form of flux scanning equipment as far more reliable information concerning reactor operation could be obtained from temperature measurements. However, such a system is ruled out on economic considerations and some form of compromise has to be adopted.

In the case of the gas-cooled reactors gags are placed at the bottom of each channel to restrict the gas flow. These gags may be varied by either entering the reactor vessel to adjust the size of the orifice or by discharging the gag and replacing with one of a different size. The main disadvantage of this system is that all alterations to the gag size must be carried out before the power phase of commissioning so that it is still possible to enter the reactor to adjust the gags. Hence the gags must be set up initially to a theoretical flux distribution. If the actual measured distribution at power differs from the theoretical estimate then it is usual to vary the amount of absorbing material in the reactor in order to shape the neutron flux to the gagging pattern.

It is very important to determine the actual flow in each channel at some suitable stage in the commissioning programme and adjust the gags at this stage to give the correct flow to suit a particular flux distribution. It is also essential to ensure that there is a reasonable coolant flow in each channel in the reactor. This is necessary in order to show that no channel is completely blocked due to some constructional error. If this did occur then on raising to power a fire would almost certainly start in the blocked channel with fuel melt-out. Hence even if accurate flow measurements were not required some form of flow test in each channel would have to be carried out.

Coolant flow tests are usually carried out either before fuel loading or immediately after the reactor has been loaded to full size.

5.2.1. Measurement of Channel Gas Flow

The actual measurements are made at amospheric pressure with air circulating through the reactor and gas circuits. It would be impractical to consider pressurising the circuits with carbon dioxide for each measurement. The flow is measured by using vane anemometers placed at the top of each channel. The rate of rotation of the vane anemometer is determined as a pulse rate from a germanium photocell. As the vanes rotate it is arranged that the rotating arms interupt a beam of light falling on the photocell. Consequently a number of pulses are produced which are counted on a scaler. The measured count rate is proportional to the air flow. Each vane anemometer is calibrated initially in order to obtain absolute flow measurements.

One of the difficulties of this method is to ensure a leak-free seal between the anemometer and the top of the gas channel. It is essential if any degree of accuracy is to be achieved to ensure that all the air flowing in the channel passes through the anemometer. For this reason the anemometers are mounted on a probe with an inflatable seal which is inserted into the channel. A typical arrangement is shown in Fig. 5.2.

Most gas-cooled reactors have a large number of channels

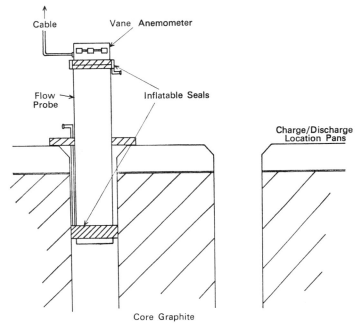

FIG. 5.2. Anemometer and flow probe

(usually in excess of 3000) and consequently it is an advantage to use as many anemometers as possible in order to reduce the time taken to obtain a complete set of results. An arrangement using 128 anemometers is shown in Fig. 5.3.

As has been mentioned, the experiments are performed using air at atmospheric pressure and ambient temperature. Under reactor operating conditions carbon dioxide would be used at pressures in excess of 100 psig and with temperatures varying between 140°C to 350°C. It is usual to apply corrections to the results obtained in order to extrapolate to actual operating conditions.

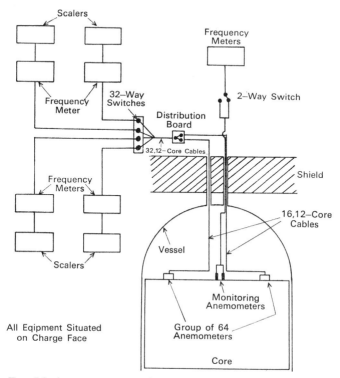

Fig. 5.3. An arrangement using 128 anemometers in two separate groups

5.3. GAS DISCHARGE FROM STACKS

The release of gaseous radioactive waste from nuclear reactor installations is important from the design and operations point of view. The amount of radioactive gas to be released into the atmosphere and the method of release may well effect the overall design of the plant and may restrict the choice of siting for the installation. Fortunately the amount of necessary discharge to the atmosphere from nuclear power plants is very small and can easily be dealt with by the suitable design of a stack to prevent undue exposure to the surrounding environment.

A certain amount of radioactive gaseous discharge to the atmosphere must take place. This arises from three main sources.

(a) *Argon-41*

In many reactors air is used to cool certain reactor components outside the reactor pressure vessel. In some cases air is actually used as the fuel coolant. In other cases the core itself is surrounded by air. Under neutron bombardment the radioactive argon-41 isotope is produced which has a 110 min half-life. The hazard from this isotope arises from the external gamma dose rate from a cloud of the gas. The dose rate produced as a result of inhalation is comparatively small. The permissible dose rate for whole body irradiation from one type of aerial effluent is 4 mrad/week.[2]

(b) *Carbon-14*

Many power reactors use carbon dioxide as coolant which circulates in a closed circuit. The neutron irradiation of carbon-13 as CO_2 produces the beta-active isotope carbon-14 with a half-life of 5600 years. In addition other radioactive materials may be present in the coolant gas. Apart from leakage from the gas circuits which is usually small, it is a requirement to be able to discharge the coolant from time to time. Consequently suitable arrangements have to be made to discharge the active carbon dioxide without causing a radiation hazard to the environment. The permissible average concentration[2] of carbon-14 in air is 200 $\mu\mu c/m^3$. This would give rise to a dose rate of 4 mrad/week.

(c) *Iodine-131*

The isotope iodine-131 which has an 8·0 day half-life is produced as a result of fission. Normally the fuel is contained and therefore there is no release of iodine-131 into the atmosphere. However, it must be borne in mind that at the time of discharge of the reactor coolant a certain amount of iodine may be present due to burst fuel elements. Burst fuel elements are of course detected at an early stage of development and removed from the reactor. However, consideration must be given to iodine-131 when contemplating the discharge of a gaseous coolant to the atmosphere. The main hazard from iodine-131 is in its transference from grass to cows' milk and then to the human thyroid. The maximum permissible level in grass is 8 mμc/m^2 of ground surface which has been calculated to give rise to a dose rate of not greater than 3 rad/year to the thyroid.[2]

The radioactive gases produced as a result of nuclear reactor operation are usually discharged to the atmosphere through a suitably designed stack.

The maximum permissible dose rate for external whole body radiation to the general public is 0·5 rad/year. However, it is usual to take a rate equal to two-fifths of this value (i.e. 0·2 rad/year) for the total dose rate from a particular gaseous discharge from one installation. This is advisable so that the site may be used for expansion and the construction of a second installation at some future date. During gaseous discharge these limits for external radiation must not be exceeded. In addition consideration must be given to risks due to inhalation and the contamination of grass and herbage. With regard to inhalation it is found that the risks due to external radiation from argon-41 and from the uptake through food chains from carbon-14 and iodine-131 are more restrictive than inhalation.

It is accepted therefore that discharge of gaseous waste is necessary and steps must be taken to ensure that this can be done in a safe manner. In the initial design of the nuclear reactor a stack will be designed of a height which will ensure that the gaseous discharge will have dispersed sufficiently by the time it reaches ground level.

A number of mathematical formulae have been developed for estimating the height of the plume of gas emitted from a stack and the resulting ground level concentration of the discharge downwind of the stack. However, as will be appreciated it is impossible to allow for all meteorological conditions in calculating these quantities and for this reason a number of empirical formulae have been developed based on wind tunnel information.[3] In any case it is always necessary to carry out some type of experimental test of the stack before proceeding to discharge active gases. These tests usually take the form of smoke bomb experiments. Stack discharge emission rates and temperatures are initially established with air as the effluent. A suitable smoke bomb is then placed at the bottom of the stack and observations carried out on the behaviour of the smoke plume. The tests are always repeated for a number of different types of weather conditions. Finally, before active discharge takes place carbon dioxide is discharged under suitable weather conditions and a carbon dioxide survey carried out in the immediate vicinity of the stack and the surrounding countryside. As a result of these tests it may be necessary to modify the stack.

5.3.1. Calculation of Ground-level Concentrations

Under ideal meteorological conditions the concentration at ground level from a source at a height h above the ground has been obtained by O. G. Sutton[4] to be:

$$X(x, y) = \frac{2S}{\pi C_y C_z U x^{2-n}} \exp\left[-\frac{1}{x^{2-n}}\left(\frac{y^2}{C_y^2} + \frac{h^2}{C_z^2}\right)\right] \quad (5.1)$$

where $X(x, y)$ is the volume concentration per m³,

S the rate of emission per sec,

h the stack height in metres,

U the wind velocity in m/sec,

C_y and C_z are Sutton's diffusion coefficients in horizontal and vertical directions,

x, y the position of measurement from foot of source at height h in metres,

n the parameter dependent upon atmospheric stability (0·25 for adiabatic and lapse conditions).

On the midline of the gas plume the value of y in the above expression is zero and the maximum concentration along this centre line will occur at x_m given by

$$\frac{\partial}{\partial x} X(x, 0) = 0$$

i.e.

$$x_m = \left(\frac{h}{C_z}\right)^{\frac{2}{2-n}}$$

Hence, by substitution, the maximum concentration with respect to horizontal distance from the source will be given by

$$X_{max} = \frac{2S}{\pi e U h^2} \cdot \frac{C_z}{C_y} \quad \text{(per m}^3\text{)} \qquad (5.2)$$

It is found that for large values of h the above ratio of Sutton's diffusion coefficients is approximately unity.

5.3.2. Calculation of the Effective Stack Height

The discharge from a stack of height h will initially continue to rise to a height which is dependent upon the emission rate from the stack and the buoyancy of the discharge.[5] Consequently the value of h in equation (5.2) should be replaced by $h+\Delta h$, where h is the physical stack height and Δh is the height of the gas plume above the stack, the value of $h+\Delta h$ being the effective stack height.

Many formulae have been developed for calculating the plume rise. At the present time none has been universally accepted and it is usual to design the stack by using several different formulae. The two most widely used formulae are the completely empirical expressions deduced by Davidson, Bryant[6, 7] and Holland.[8, 9]

The Davidson–Bryant expression, based on wind tunnel data, is given as

$$\Delta h = \left(\frac{E_s}{U}\right)^{1\cdot4} d\left(1+\frac{\Delta T}{T_s}\right) \tag{5.3}$$

where E_s is the stack effluent velocity in m/sec,
$\quad U$ the horizontal wind velocity in m/sec,
$\quad d$ the internal diameter of stack in metres,
$\quad \Delta T$ the temperature excess of effluent over ambient air (°K), and
$\quad T_s$ the effluent temperature (°K).

This formula applies to stacks with large h.

The Holland formula is given by

$$\Delta h = 1\cdot5\,\frac{E_s}{U}\,d\left[1+1\cdot8\,\frac{Cp_e}{Cp_a}\cdot\frac{P_a}{P_0}\cdot\frac{m_e}{m_a}\cdot\frac{\Delta T}{T_s}\,d\right] \tag{5.4}$$

where Cp_e is the specific heat of effluent at constant pressure,
$\quad Cp_a$ the specific heat of ambient air at constant pressure,

P_a the ambient pressure,

P_0 the standard pressure (760 mm mercury),

m_e the molecular weight of effluent, and

m_a the molecular weight of air.

5.3.3. Conditions affecting the Plume Rise

It must be pointed out that the observed plume height is often widely different from that calculated by the two expressions given in Section 5.3.2. This is because many factors effect the rise of the plume and the experimental data available has been obtained with differing meteorological conditions making comparisons difficult. However, the expressions do tend to underestimate the plume rise in ideal conditions which adds a safety factor as far as a radiation hazard is concerned.

The plume rise and hence the distributed ground level concentration are affected by surrounding buildings, the shape of the stack, the surrounding countryside and meteorological conditions.

(a) *Surrounding buildings*

Buildings and other obstructions to air flow will effect the rise of the plume from a stack.[10, 11] These effects are not easy to calculate and the problem is usually tackled by carrying out a number of wind tunnel tests. The effect of nearby buildings can be reduced by increasing the emission velocity of the discharge from the stack.

An empirical rule has been deduced from available data which states that the effect of surrounding buildings is minimised if the stack is at least $2\frac{1}{2}$ times the height of the nearest building.

(b) *Shape of the stack*

The height of the gas plume varies to some extent with the shape of the stack. From the results of wind tunnel experiments it has been found that greatest plume rise is obtained with a rectangular stack positioned in the direction of the wind followed by a circular stack, a square stack and a rectangular stack positioned perpendicular to the wind.

(c) *Surrounding countryside*

Air flow is greatly effected by mountains, hills, trees, large areas of water, roads and rivers. For instance, there may be large concentrations of discharged effluent on the leeward side of a ridge if the discharge takes place on the windward side even with large separation distances. Wind tunnel experiments and other tests at the site should be carried out for all but extremely uniform terrains.

(d) *Meteorological effects*

The behaviour of a plume of discharge emitted from a stack is obviously extremely dependent upon local meteorological conditions.[12, 13, 14] The two factors which cause most influence on plume behaviour are wind and air stability. An accurate relationship between plume rise and wind speed has not yet been determined. Equation (5.3) shows that the effective stack height is dependent upon $(1/U)^{1 \cdot 4}$ whereas equation (5.4) gives a $1/U$ variation. Stability of the atmosphere is dependent upon wind speed and upon the vertical temperature gradient. Obviously with very strong winds a large amount of mixing occurs and the behaviour of the plume is influenced mostly by the character of the wind. With light winds the behaviour of the plume is most influenced by the variation of temperature with height.[15]

The effect of vertical temperature gradients on the behaviour of a plume under light wind conditions is illustrated in Fig. 5.4. The looping plume illustrated in (a) is typical of normal daytime conditions with clear skies when the temperature decreases with height. The coning plume occurs with higher wind conditions and only a slight decrease in temperature with height. Fanning occurs at night, just before sunset and after

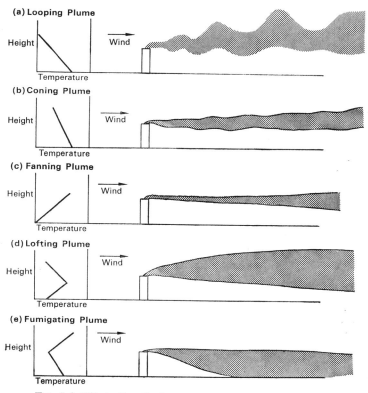

Fig. 5.4. Effect of vertical temperature gradient on plume behaviour

sunrise. It is produced as a result of an increase in temperature with height and represents an inversion. When an inversion is building up, a condition exists when the temperature increases with height near to the ground surface and then above a certain height decreases again. This gives rise to a lofting plume illustrated in (d). It occurs normally in the late afternoon or early evening. Finally a fumigating plume occurs when an inversion is being destroyed and the temperature decreases with height near to the ground surface but increases above a certain height. The conditions occur in the early morning.

In view of the uncertainties associated with the behaviour of the gaseous discharge emitted from stacks it is essential during the commissioning stages of reactor operation to carry out a number of tests to verify the effectiveness of the designed stack. From time to time it is necessary to discharge the coolant completely from the reactor vessel. In this case it is usual to wait for good meteorological conditions before commencing the discharge. The lofting plume in Fig. 5.4(d) which occurs in the late afternoon gives ideal conditions for discharge. Due to the nature of the plume the lowest ground concentrations occur in this case.

REFERENCES

1. Manual for the operation of research reactors. *I.A.E.A. Technical Reports Series* No. 37 (1965).
2. *Recommendations of the Int. Comm. of Radiological Protection.* Pergamon (1959).
3. BOSANQUET, C. H. and PEARSON, J. L. The spread of smoke and gases from chimneys. *Trans. Faraday Soc.* **32,** 1249 (1936).
4. SUTTON, O. G. The problem of diffusion in the lower atmosphere. *Quart. Journ. Royal Meteorological Soc.* **73,** 257 (1947).
5. GIFFORD, F. A. Atmospheric dispersion. *Nuclear Safety* **1** (3), 56 (1960).
6. BRYANT, L. W. and COWDREY, C. F. *Proc. Inst. Mech. Engrs.* **169,** 371 (1955).
7. DAVIDSON, W. F. The dispersion and spreading of gases and dust from chimneys. *Trans. Conf. on Industrial Wastes,* American Annual Meeting Industrial Hygiene (1954).

8. HOLLAND, J. Z. A meteorological survey of the Oak Ridge area. *U.S.A.E.C. Report* ORO-99 (1953).
9. LUCAS, D. H., MOORE, D. J. and SPUR, G. The rise of hot plumes from chimneys. *Int. J. Air–Water Pollution* **7**, 473 (1963).
10. SUTTON, O. G. *J. Inst. of Fuels* **33**, 495 (1960).
11. LUCAS, D. H. Symposium on the dispersion of chimney gases. *Int. Journ. Air–Water Pollution* **6**, 94 (1962).
12. DAVIDSON, B. Turbulence and wind variability observations. *J. Appl. Met.* **2** (4), 463 (1963).
13. U.S. WEATHER BUREAU. Meteorological and atomic energy. *U.S.A.E.C. Report*, A.E.C.U. 3066 (1955).
14. MOSES, H. *et al*. Effects of meteorological and engineering factors on stack plume rise. *Nuclear Safety* **6**, 1 (1964).
15. GLUECKAUF, E. (Ed.). *Atomic Energy Waste*. Butterworths (1961).

CHAPTER 6

Reactor Physics Experiments

ON THE satisfactory completion of the pre-nuclear tests, the nuclear phase of commissioning a reactor may be commenced. As in the case of the conventional tests, the aim of the nuclear experiments is to enable the operating staff to obtain reliable information about the reactor core and its components prior to carrying out any full-scale operation of the reactor. Planning of the nuclear phase both as regards the type and number of experiments to be carried out and the order in which these tests are performed is vitally important. Nuclear commissioning should proceed with care and the result of each experiment should be analysed and discussed by the operating and design staffs before proceeding to the next stage. It is sometimes necessary to reconsider the complete programme if some unexpected result is obtained during a particular experiment. The nuclear tests should be related as closely as possible to the normal operating state of the reactor. If it is not possible to test the normal operational arrangement, then it may be necessary to carry out a series of tests, the purpose of which is to enable reasonably accurate extrapolation to be made to the normal state.

The objects of the nuclear commissioning experiments are to obtain information which will enable the reactor and associated plant to be brought into operation in a safe and reliable manner, to obtain information to check the design data and to provide information which will enable future design assessments to be made with greater accuracy.[1, 15] Advance theoretical and design information is only used by the operating staff to provide a basis for the planning of a commissioning programme. Theory is used as a guide in reactor operation and not as a principle.

Individual commissioning tests must be dependent to some extent on the type, use, and characteristics of a particular reactor. For all reactors the nuclear phase programme commences with the initial loading of nuclear fuel into the core, and is completed with the raising of the reactor to power for the first time. The programme invariably includes:

1. Loading to criticality.
2. Reactor parameter measurements.
3. Loading to full size.
4. Power measurements.
5. Coolant flow tests.
6. Control rod calibrations.
7. Flux distribution measurements.
8. Measurement of shut-down capacities.
9. Temperature coefficient measurements.
10. Reactivity effects (absorbers, voids, loss of moderator, etc.).
11. Burst element detection gear tests.

The individual experiments will be considered in more detail in this and other chapters.

6.1. LOADING TO CRITICALITY

Fuel is always loaded initially into a reactor core in a very careful and controlled manner. The critical core loading, which is the amount of fuel required to enable a self-sustained chain reaction to occur, is approached gradually and from the measurements carried out during this approach an accurate estimate is made of the critical loading (i.e. critical size or critical mass).[2, 3, 4, 15]

An approach to criticality is carried out by measuring the neutron flux at several positions in or near the reactor core as fuel is loaded into the core in a uniform manner.

Let us consider the relationship between the neutron flux and the core lattice parameters. Initially there is always a source of neutrons which is either provided by means of a radioactive neutron source or is a result of spontaneous fission within the reactor fuel itself.

The source neutrons produced in the core are multiplied due to fissions occuring in the reactor fuel.[5] The multiplication M is defined as

$$M = \frac{\text{total thermal neutron flux (source + fission)}}{\text{thermal neutron flux due to source}}$$

Hence the degree of multiplication is dependent upon the effective multiplication factor (k_{eff}) of the reactor core.

If a source of neutrons is introduced into a subcritical assembly of fuel such that their energy spectrum and spatial distribution are characteristic of the assembly, then the multiplication is the total number of neutrons appearing in the fissile material per source neutron. Assuming T source neutrons there will be Tk_{eff} neutrons after the first fission generation, Tk_{eff}^2 neutrons after the second fission generation, Tk_{eff}^3 after the third generation and so on. Hence the total number of neutrons per source neutron is given by

$$M = \frac{T + Tk_{\text{eff}} + Tk_{\text{eff}}^2 + Tk_{\text{eff}}^3}{T}$$

$$= 1 + k_{\text{eff}} + k_{\text{eff}}^2 + k$$

After many generations, the multiplication M approaches

$$M = \frac{1}{(1 - k_{\text{eff}})} \qquad (6.1)$$

for $k_{\text{eff}} < 1$.

At criticality $k_{\text{eff}} = 1$ by definition and the multiplication M is infinite and a self-sustaining chain reaction is produced.

Using a simple treatment for a large reactor the relationship between the effective multiplication factor and the infinite multiplication is given by

$$k_{\text{eff}} = \frac{k_\infty}{1 + \alpha^2 M_z^2 + \beta^2 M_r^2} \qquad (6.2)$$

where $M_r^2 = L_r^2 + L_{sr}^2$,

$\quad M_z^2 = L_z^2 + L_{sz}^2$,

$\quad \beta \quad$ is the radial Laplacian,

$\quad \alpha \quad$ the axial Laplacian,

$\quad M_r^2 \quad$ the radial migration area,

$\quad M_z^2 \quad$ the axial migration area,

$\quad L_r \quad$ the radial diffusion length,

$\quad L_z \quad$ the axial diffusion length,

$\quad L_{sr} \quad$ the radial slowing down length, and

$\quad L_{sz} \quad$ the axial slowing down length.

If $k_{\text{eff}} = 1$, then

$$k_\infty = 1 + \beta_c^2 M_r^2 + \alpha_c^2 M_z^2$$

β_c and α_c being the values of the Laplacians at criticality)

and

$$\frac{1 - k_{\text{eff}}}{k_{\text{eff}}} = \frac{(1 + \beta^2 M_r^2 + \alpha^2 M_z^2 - 1 - \beta_c^2 M_r^2 - \alpha_c^2 M_z^2}{k_\infty}$$

$$= \frac{M_r^2(\beta^2 - \beta_c^2) + M_z^2(\alpha^2 - \alpha_c^2)}{k_\infty}$$

If the reactor fuel is loaded to maintain radial symmetry whilst keeping the axial geometry constant then

$$\frac{1 - k_{\text{eff}}}{k_{\text{eff}}} = \frac{M_r^2(\beta^2 - \beta_c^2)}{k_\infty} \qquad (6.3)$$

Let N_s be the measured value of the thermal neutron flux due to the source term alone and N_t the total thermal flux due to source and fission, then

$$M = \frac{N_t}{N_s}$$

Using equations (6.1) and (6.3) we get

$$\frac{N_s}{N_t} = \frac{M_r^2(\beta^2 - \beta_c^2)}{k_\infty} \qquad (6.4)$$

at the near critical condition.

The value of β^2 is dependent upon the geometrical shape of the reactor core. For instance:

$$\beta^2 = \left(\frac{\pi}{R_e}\right)^2$$

for a spherical core where R_e is the extrapolated radius

and
$$\beta^2 = \left(\frac{j_0}{R_e}\right)^2$$

for a cylindrical core.

Similarly corresponding values of α^2 are given by

$$\alpha^2 = \left(\frac{\pi}{R_e}\right)^2 \quad \text{and} \quad \alpha^2 = \left(\frac{\pi}{H_e}\right)^2$$

where H_e is the extrapolated height of the core.

Hence, in the case of a cylindrical core, equation (6.4) becomes

$$\frac{N_s}{N_t} = \frac{M_r^2 j_0^2}{k_\infty} \left[\frac{1}{R_e^2} - \frac{1}{R_{ec}^2} \right] \qquad (6.5)$$

where R_{ec} is the value of the extrapolated radius at criticality. As M_r^2, j_0^2, k_∞, N_s and R_{ec}^2 are constants, then

$$\frac{R_e^2}{N_t} \quad \text{is proportional to} \quad R_{ec}^2 - R_e^2 \qquad (6.6)$$

A similar expression can be deduced for cores with different geometries.

The mass of fuel loaded into the core is proportional to the square of the radius of the core, and so

$$\frac{M}{N_t} \quad \text{is proportional to} \quad M_c - M$$

where M = mass of fuel loaded,
M_c = mass at criticality.

An approach to criticality and the determination of the critical mass is carried out by measuring the thermal neutron flux in the core with different increasing amounts of fuel loaded.[6, 7] M/N_t is then plotted against M and this approach to criticality curve, as it is called, is extrapolated to cut the M axis at the point where M/N_t is zero.

Before commencing to load fuel into a reactor it is necessary to consider three factors. These are:

1. The position of the installed radioactive start-up source (if used).
2. The positions at which the thermal neutron flux is to be measured.

3. The initial amount of fuel to be loaded for the first measurements.

The neutron flux measurements are usually made using boron trifluoride proportional counters placed at several positions in or near the reactor core. Obviously approach curves should be obtained using more than one set of measurements and it is recommended that at least three counters are used.

To maintain clean conditions in the reactor core, the counters and associated plugs are usually enclosed in polythene covers. The counters may be placed in the top reflector of the core or secured to the sides or top of the core moderator or reflector. In the case of a large gas-cooled graphite reactor, it is often possible to place a counter at the physical centre of the core by lowering it down a central channel, hence giving an extremely good approach curve.

A radioactive start-up source is nearly always used (see Chapter 3) to provide an initial neutron flux. The position of the flux measuring boron trifluoride counters with relation to this source is extremely important. As fuel is loaded neutrons emitted by the source must be multiplied by the fuel and then measured. It is usual to arrange the counters in positions such that as fuel is loaded the fuel screens the neutron source from the counters.

The change in the geometrical shape of the core as fuel is loaded will also effect the approach to criticality curve. During loading every effort should be made to load the fuel as uniformly as possible and to maintain the same geometrical shape. In the case of a large natural uranium power reactor, it is quite easy to satisfy the requirements of good geometry and good counter positioning. It is not so easy in the case of the smaller enriched uranium cores however. In this latter case it is normally necessary to load fuel elements which are large with respect to the overall size of the core and so the geometry changes radically. Hence, approach curves are not straight lines in the early stages.

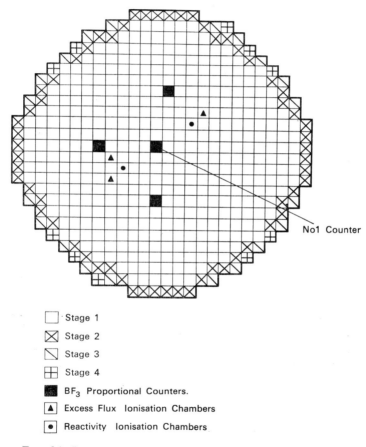

No1 Counter

☐ Stage 1

☒ Stage 2

◻ Stage 3

⊞ Stage 4

■ BF₃ Proportional Counters.

▲ Excess Flux Ionisation Chambers

● Reactivity Ionisation Chambers

FIG. 6.1. Instrument positions and loading pattern for a large cylindrical reactor

Reactor Operation

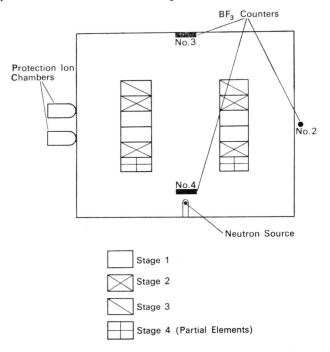

Fig. 6.2. Instrument positions and loading pattern for an Argonaut type research reactor

Typical arrangements for source and counters for two completely different cores are shown in Figs. 6.1 and 6.2.

A typical large power reactor is shown in Fig. 6.1 and it will be noted that it is quite easy to maintain a cylindrical shape throughout the complete loading. Figure 6.2 shows a research reactor. In this case a constant geometrical shape is only attained for the last few loading stages.

Figure 6.3 indicates typical approach curves which may be obtained. Curve 1 is for a large natural uranium reactor with readings obtained from a counter in position 1 on Fig. 6.1.

Curves 2 and 3 are for a research reactor with readings obtained from counters 2 and 3 in Fig. 6.2. Note that in this case a straight line is not obtained until the last few loading stages. Curve 4 would be obtained from counter 4 (Fig. 6.2). This is a very bad approach and an extremely dangerous one as in the early stage of loading extrapolation would lead to a large overestimate of the actual critical size. The curve is obtained because most of the thermal flux at that particular position is due to the source neutrons alone.

Immediately the counters and neutron source have been installed a source background flux measurement is carried out with no fuel loaded in the core. The reactor is then ready for the first stage of loading. A decision must now be made regarding the amounts of fuel to be loaded for the first two

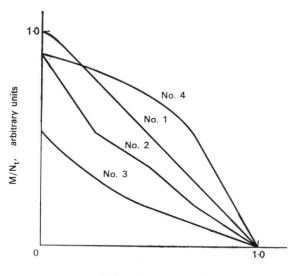

Fig. 6.3. Typical approach to criticality graphs

stages. Until measurements have been obtained with two different amounts of fuel loaded, it is not possible to estimate a value for the critical size. Various criteria have been adopted for determining the initial loading, but it is usual nowadays to load one-half or one-third of the theoretically estimated value. In the case of a new reactor core with no experimental information available from similar types, one-third is usually taken, in all other cases one-half is loaded.

Table 6.1 gives actual examples of initial stages for fuel loading.

TABLE 6.1. INITIAL FUEL-LOADING STAGES

Reactor	Calder Hall	Berkeley	Queen Mary College Critical Assembly
Estimated value	470 channels	654 channels	3·0 kg U-235
First stage	240 channels	298 channels	1·083 kg U-235
Second stage	300 channels	396 channels	1·626 kg U-235
Measured value	406 channels	744 channels	3·212 kg U-235

For the second stage of loading it is usual to load approximately one-half of the amount loaded for the first stage. However, the actual amount will to some extent depend on the geometry of the core as much as anything else.

After obtaining neutron flux measurements for the first two stages, M/N_t against M is plotted for all the counters. A straight line is drawn through the two points for each counter and extrapolated to cut the M axis at M/N_t zero. Different values of the critical mass will probably be obtained for each counter and the minimum value is the first estimate of the minimum critical mass. For the third stage half the difference between the estimated minimum critical mass and the amount loaded at stage 2 will be loaded. The approach to criticality curve is continued and for stage 4 half the difference between the amount loaded at stage 3 and a new estimated minimum

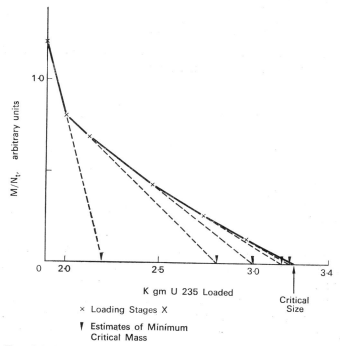

Fig. 6.4. Approach to criticality graphs and estimate of minimum critical size at different loading stages

critical mass is loaded. This procedure is continued and the estimated value of the minimum critical mass obtained after each stage. The process is terminated before criticality is achieved. This end point is dependent upon the level of the neutron flux measured at the final stages. For instance, at

Calder Hall the approach was terminated 6 channels from criticality, and at Queen Mary College 10 g from criticality.

The approach curves for all counters should converge at the critical point during the final stages. Figure 6.4 represents typical curves obtained during an actual approach for the Queen Mary College critical assembly.

The broken lines show how the estimates of the minimum critical mass vary during the approach.

6.2. SAFETY ASPECTS OF AN APPROACH TO CRITICALITY

An approach to criticality is carried out with the control rod system completely withdrawn from the reactor core. Consequently certain safety precautions should always be adopted during the measurements. During the physical loading of the core the control rods are always fully inserted. In the case of water reactors in which the water moderator is dumped from the core in the normal shut-down state, then the water would be dumped during the loading. In other words, fuel loading is only carried out with the reactor shut-down system fully operative. In addition all normal reactor safety systems (i.e. automatic scram and shut-down devices and the nuclear instrumentation) must be fully operational.

If the normally installed instrumentation is situated some distance from the core in an external thermal column it would be quite inadequate for protection during an approach to criticality. The neutron flux at these instruments would be far too low to be detected. The problem is solved by placing temporary ionisation chambers on the top surface of the moderator, or near to the area of the core which is to be loaded initially. These temporary instruments are then connected to the normal flux level and period safety channels, which are suitably set to provide adequate low level protection.

After each stage of loading the fuel, the control rod system is withdrawn completely from the core and the thermal flux measurements obtained. The rods are fully inserted again before loading the next stage.

It is usual to carry out additional measurements during the approach to give an indication of the effectiveness of the control rod system. Therefore, at each loading stage, neutron flux measurements are determined with a control rod or combination of rods inserted by varying amounts. Approach curves are plotted for each case considered and estimates made of the critical mass with the rods inserted. From these values estimates may be made of the worth of the rods and hence loading of fuel may be continued with the knowledge that the control rod system gives adequate protection.

6.3. REACTOR PARAMETER MEASUREMENTS

It is usual, following the measurement of the initial critical size of a reactor core, to carry out several reactor physics measurements on the "just critical" reactor.[7, 8, 9, 15] These measurements, which are not necessarily desirable from an operational viewpoint, provide information concerning the basic nuclear data of the reactor. The measurements are, of course, of great interest and of use to the designers as it enables checks on the accuracy of core calculations to be made.

The extent and importance of the reactor physics measurements is very dependent upon the type and eventual use of the reactor. In the case of a large power station, the tendency is to reduce the physics measurements to a minimum so that the station is brought to power as quickly as possible.

In the case of research reactors, a thorough experimental investigation of the behaviour of the core and the measurement of the core parameters is highly desirable. In addition the accuracy and type of the experiments carried out is dependent upon the type of core. In large reactors with a multiplicity of regions there is a different set of parameters for each region, and measurement of anything but basic data is difficult, laborious and inaccurate.

In small reactors using enriched uranium, it is difficult to add fuel in sufficiently small amounts, or to change the geometry in a sufficiently uniform manner to be able to make extensive and accurate parameter measurements.

However, whatever type of reactor is being commissioned, it is usual to carry out a basic minimum number of reactor physics measurements. The experiments usually performed are:

1. The measurement of the variation of the effective multiplication factor with fuel loading, and the determination of the fuel coefficient of reactivity.
2. The measurement of the thermal neutron flux distributions in the horizontal and vertical planes of the core.

6.3.1. Variation of $k_{\text{effective}}$ with Fuel Loaded above Criticality

The equation which relates the reactivity or excess reactivity of a supercritical reactor to the nuclear properties of the reactor is expressed as

$$\varrho = \frac{l}{Tk_{\text{eff}}} + \sum_{i=1}^{m} \frac{\beta_i}{1 + \lambda_i T} \tag{6.7}$$

where excess reactivity ϱ is defined as

$$\varrho = \frac{k_{\text{eff}} - 1}{k_{\text{eff}}}$$

where l is the prompt neutron lifetime,

T the stable reactor period,

β_i the fraction of total number of fission neutrons which are delayed neutrons of the ith delayed group, and

λ_i the radioactive decay constant of the ith precursor.

Hence, a relationship is provided between the reactivity of a core and the stable period.[2] Since for a particular reactor values of l, and of β and λ for the various groups of delayed neutrons may be calculated, then if the reactor period T is known, the reactivity ϱ may be calculated. The above equation is known as the inhour equation and is used extensively in the analysis of reactor physics measurements.

By using the inhour equation it is possible to plot the relationship between reactivity and reactor period for any reactor. It is, however, more useful to establish a reactivity scale in terms of the reactor doubling time rather than the

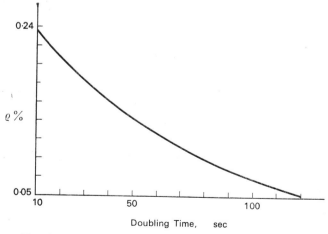

FIG. 6.5. Typical doubling time–reactivity relationship

e-folding time (reactor period). As the doubling time T_d is given by

$$T_d = T \log_e 2$$

then the doubling time relationship may easily be established.

Figure 6.5 is a typical doubling time–reactivity relationship for a water moderated enriched uranium reactor.

Experimental procedure

On completing the approach to criticality measurements, fuel is loaded in the reactor core as uniformly as possible to increase the core loading above the critical size. At the end of the loading, the control rods will be completely withdrawn and a reactor doubling time measured. At the end of the doubling time measurement the control rods are tripped into the core to shut the reactor down. Caution must be exercised in determining the initial amount of fuel to be loaded above criticality. Obviously a very short doubling should not occur and it is usual to attempt to estimate the fuel coefficient of reactivity (i.e. reactivity change per unit mass of fuel) from the results obtained in determining the critical size and the design data.

Having obtained an estimate, then enough fuel should be loaded above the measured critical size to give a doubling of not less than 100 sec. If the estimate of the fuel coefficient is not considered to be very reliable then approximately one-half of the estimated amount of fuel should be loaded for the first measurement.

After the first measurement more fuel is loaded and a second doubling time determination carried out. The procedure is repeated and several doubling times over the range 100 to 15 sec may be obtained. A plot of reactivity against fuel loaded above criticality should be made and the slope of this plot gives the fuel coefficient of reactivity.

6.3.2. The Measurement of Doubling Times

Any instrument which measures the thermal neutron flux in or near the reactor core may be used for doubling time determination. In the core of small reactors the normally installed instrumentation may be used. In the case of large reactors, additional ionisation chambers are positioned near to the core and are connected to suitable linear current measuring instruments. These instruments can either be normally installed instrumentation in the control room, or some special instrument adapted for this purpose.

As in the measurements during the approach to criticality, instruments are required for protection during doubling time measurements.

In the case of small reactors, the normally installed instrumentation placed immediately adjacent to the core will receive adequate neutron flux for measurement.

In the case of a large power reactor, temporary instruments have to be installed near to the core. These instruments are connected into the normal excess flux and period protection circuits to provide suitable protection.

Figure 6.6 indicates a typical arrangement which is necessary for a large gas-cooled natural uranium reactor in order to carry out reactor physics measurements.

It is essential to measure doubling times over a large number of decades to achieve accuracy. A stable reactor period is not established until the reactor flux has doubled 3 or 4 times so that the initial transients have died away. It is usual to measure the time for the flux to double over 5 or 6 decades and then obtain an average doubling time from the last 2 or 3 decades only.

At these early stages in the life of the reactor the total power must not be allowed to rise too high during these measurements to prevent appreciable activation of the core.

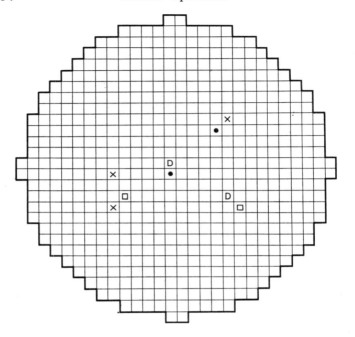

☒ Excess Flux Protection

⦿ Low Power Measurement

☐ Doubling Time Measurement

▣ Reactivity Protection

FIG. 6.6. Typical arrangement of temporary instruments for
reactor physics measurements in a large reactor

6.3.3. Measurement of the Effects of Absorbers and Voids

During the measurements of the variation of k_{eff} with
fuel loaded above criticality, it is convenient to carry out
reactivity measurements associated with the addition of ab-

sorbers to the reactor core.[7, 10, 15] These measurements are very dependent upon the type of reactor which is being commissioned and the absorbers in question. For instance, in the case of graphite moderated gas cooled natural uranium reactors, the critical size will have been determined with air in the pressure vessel. During normal operation the vessel will be filled with carbon dioxide which is a non-neutron absorber relative to air. Consequently in order to be able to deduce the value of the critical size in carbon dioxide, an air pressure coefficient of reactivity must be determined.

This is usually carried out at a certain stage during the loading above criticality. For instance, after a doubling time measurement of 100 sec has been carried out, the control rods are inserted and the air pressure within the vessel is reduced by approximately 5 cm of mercury. The control rods are withdrawn and a second doubling time measured. This will be repeated with a further reduction of 5 cm of mercury in the air pressure. From the results an air pressure coefficient of reactivity may be evaluated. Typical results are given in Table

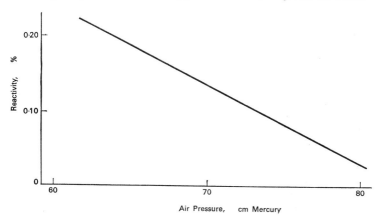

FIG. 6.7. Reactivity–air pressure graph

TABLE 6.2. VARIATION OF REACTIVITY
WITH AIR PRESSURE

Pressure (cm of mercury)	Doubling time (sec)	Reactivity (%)
76·9	100	0·060
70·8	34·4	0·124
67·6	20·7	0·173

6.2 and a plot of reactivity against pressure shown in Fig. 6.7. The slope of this graph gives the air pressure coefficient in terms of reactivity per cm of mercury pressure.

In the case of smaller reactors and water moderated reactors, the effect of adding a uniformly distributed absorber throughout the core may be investigated. This would be carried out with a core loading above criticality giving a doubling time of approximately 20 sec. Uniformly distributed absorber would be placed in the core in two stages and doubling times measured at each stage. From the results an absorber coefficient of reactivity (i.e. reactivity change per unit amount of absorber) would be determined.

As will be seen later (Chapter 7) the uniform absorber method (air for gas-cooled reactors and material for water reactors) is one of the best methods for control rod calibrations and hence these initial measurements with the "just critical" core are of importance.

Void coefficients (i.e. reactivity per unit void) may also be measured by similar techniques during the initial reactivity measurements.[11, 12, 15] In large power reactors, the reactivity effect of removing a channel of fuel is of great importance, especially if on-load fuel charging and discharging is to be employed. Again at a suitable stage in the "just critical" core measurements, doubling times may be measured with and without a channel filled with fuel.

The presence of air or steam voids in water reactors is important from an operational point of view and an initial determination of the effect on reactivity of such a void can be determined by the methods described.

6.3.4. Flux Distribution Measurements

Several flux distribution measurements are usually carried out with the "just critical" reactor. The measurement of the flux in the horizontal and vertical planes of the core and in the neighbourhood of the fuel provides information which enables design calculations to be checked and also provides information useful for the future operation of the reactor. For instance in a power reactor the shape of the flux distribution near to a fuel element and across the gap between two fuel elements will give an indication of the possible location of hot spots in the fuel. These may be so severe that the total heat output of the reactor is limited by these variations in flux. In research reactors knowledge of flux distributions and the absolute magnitude of the flux in the experimental facilities is required.

Other useful information is the flux distribution across an inserted control rod, across a void and in the core moderator. Flux distributions are also used to enable the power developed in the reactor to be determined.

Flux distributions are measured by distributing suitable foils throughout the reactor, irradiating the foils and then measuring the induced radioactivity on the foil. The induced activity is a measure of the neutron flux in which the foil has been irradiated.[13, 14, 15]

A variety of materials may be used for the measurement of

flux. The requirements of foils and the materials used for flux measurements are:

1. Foils must be thin, so that the flux at the point at which it is being measured is not appreciably effected by the foil itself.
2. The foil must have uniform concentration throughout.
3. The foil must be physically manageable.
4. The material must have a fairly high neutron absorption cross-section, which should be known accurately.
5. The half-life of the product nucleus must not be too short, otherwise the induced activity will decay before and during the activity measurements.
6. The half-life of the product nucleus must not be too long, otherwise the foils will have to be irradiated for long periods to build up sufficient induced activity.

For thermal flux measurements the most usually used materials are given in Table 6.3.

TABLE 6.3. PROPERTIES OF FOIL MATERIALS

Material	Relative abundance	Neutron absorption cross-section (barns)	Half-life of product nucleus	Decay particles
Manganese	100% Mn-55	12·6	2·59 hr	β^- 2·9 MeV (max) γ 3·0 MeV (max)
Indium	95·8% In-115	145	54·3 min	β^- 1·02 MeV (max) γ 1·27 MeV (max)
Gold	100% Au-197	95	2·7 days	β^- 1·38 MeV (max) γ 1·09 MeV (max)
Tungsten	28·7% W-186	34	24 hr	β^- 1·33 MeV (max) γ 0·20 MeV (max)

The most commonly used of all these materials are manganese and gold. Foils are usually made in either square or circular form and are of approximately 1 cm or $\frac{1}{2}$ cm square or circular with a diameter of 1 or $\frac{1}{2}$ cm. The thickness of the foil is usually of the order of 0·005 in. or 0·006 in. Foil holders of perspex or polythene are used to mount foils in a reactor core. The holders have a number of designs which are pertinent to a particular reactor.

Let us consider the process of activation of a foil placed in a reactor.

Let ϕ be the neutron flux (cm^{-2} sec^{-1}),
 v the volume of foil (cm^3),
 ϱ the foil material density (g cm^{-3}),
 η the natural abundance of the isotope to be activated,
 A the atomic weight,
 ε_a the macroscopic activation cross-section,
 σ_a the neutron absorption cross-section (cm^{-1}),
 λ the decay constant of product nucleus (sec^{-1}),
 N_0 Avogadro's number,
 T_1 the irradiation time (sec),
 T_2 the decay time (sec), and
 T_3 the counting time (sec).

Number of nuclei per cm^3

$$= N = \frac{N_0}{A}\varrho$$

Number of neutrons absorbed per cm^3 per second

$$= \varepsilon_a \phi$$
$$= N\sigma_a \phi$$

Number of neutrons absorbed by foil per second

$$= \eta v \frac{N_0}{A} \varrho \sigma_a \phi$$

$$= \text{number of product nuclei formed}$$

$$= K_1 \phi$$

where

$$K_1 = \eta v \frac{N_0}{A} \varrho \sigma_a$$

At time t sec, let us assume that there are n radioactive nuclei present in the foil. Then in δt seconds there will be $K_1 \phi \delta t$ produced and $n\lambda \delta t$ lost by radioactive decay.

Increase in the number of radioactive nuclei

$$\delta n = K_1 \phi \delta t - n\lambda \delta t$$

Rate of increase of radioactive nuclei

$$\frac{dn}{dt} = K_1 \phi - n\lambda$$

Integrating and noting that

at $\qquad t = 0, \qquad n = 0 \quad$ and at $\quad t = T_1, \qquad n = N_{T_1}$

we get $\quad -\dfrac{1}{\lambda} \log_e (K_1 \phi - N_{T_1} \lambda) + \dfrac{1}{\lambda} \log_e K_1 \phi = T_1$

or

$$N_{T_1} = \frac{K_1 \phi}{\lambda} (1 - e^{-\lambda T_1}) \tag{6.8}$$

The number of radioactive nuclei present at T_1 is N_{T_1}.
The number of radioactive nuclei present at T_2 is N_{T_2},

where

$$N_{T_2} = N_{T_1} e^{-\lambda T_2}$$

Rate of decay at T_2 is

$$-\left(\frac{dN_{T_2}}{dt}\right) = \lambda N_{T_1} e^{-\lambda T_2}$$

i.e.
$$-\left(\frac{dN_{T_2}}{dt}\right) = K_1\phi(1 - e^{-\lambda T_1})e^{-\lambda T_2} \qquad (6.9)$$

The induced activity on the foil is usually measured with a geiger counter or scintillation counter and associated counting equipment. A correction must be applied to the mean count rate as derived from the ratio of the total count to the counting period.

Mean count rate
$$\left(\overline{\frac{dc}{dt}}\right) = \frac{\int_0^{T_3}\left(\frac{dc}{dt}\right)_t dt}{\int_0^{T_3} dt}$$

i.e.
$$\left(\overline{\frac{dc}{dt}}\right) = \frac{1}{T_3}\int_0^{T_3}\left(\frac{dc}{dt}\right)_0 e^{-\lambda t}\, dt$$

$$= \frac{1}{T_3}\left(\frac{dc}{dt}\right)_0 \frac{1}{\lambda}\left[1 - e^{-\lambda T_3}\right]$$

and $\quad \left(\dfrac{dc}{dt}\right)_0 = \dfrac{dN_{T_2}}{dt} = $ rate of decay at the start of counting.

Then $\left(\overline{\dfrac{dc}{dt}}\right) = \dfrac{1}{\lambda T_3}(1 - e^{-\lambda T_3})K_1\phi(1 - e^{-\lambda T_1})e^{-\lambda T_2}$

and so $\quad \phi = \left(\overline{\dfrac{dc}{dt}}\right)\dfrac{\lambda T_3}{K_1(1 - e^{-\lambda T_3})(1 - e^{-\lambda T_1})e^{-\lambda T_2}} \qquad (6.10)$

The value of $\left(\overline{\dfrac{dc}{dt}}\right)$ is the value of the mean absolute rate of decay of foil activity. However, in practice the efficiency of the counting equipment must be taken into account for absolute accuracy. If K_2 is defined as the efficiency factor for the counting equipment, then the expression for the neutron flux is obtai nedby multiplying the value of ϕ in equation (6.10) by the factor K_2.

Many varieties of counting equipment are available, all being versions of a geiger counter or a scintillation counter

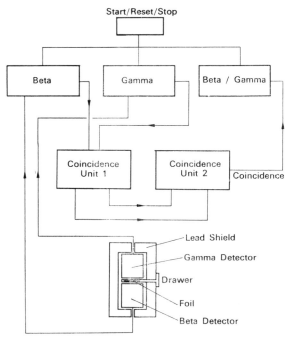

FIG. 6.8. Typical foil-counting arrangement

mounted in a lead castle and provided with a slider mechanism with which it is possible to place the active foil close to the counter. The number of counts obtained is measured using some type of scaling equipment. A typical counting arrangement is indicated schematically in Fig. 6.8.

This system incorporates two scintillation detectors on opposite sides of a foil carrier enclosed in a lead castle. One scintillation detector is beta sensitive and the other gamma sensitive. As shown in the figure the beta and gamma activity of a foil placed on the carrier may be determined separately on two different scalers. A coincidence unit is used in order to obtain a count on a third scaler when both beta and gamma rays are emitted simultaneously. A single stop (reset) start button is arranged to operate on all three scalers at the same time.

The counting of small numbers of irradiated foils with a single arrangement of counting equipment is fairly straight-forward. However, if large numbers of foils have to be counted, then it will be necessary in order to avoid lengthy periods of counting and associated fall off in foil activity, to use several sets of counting equipment. Flux distribution measurements at a large nuclear power station may well involve the irradiation of 150 to 200 foils at any one time. If large numbers of foils are irradiated then the differences associated with individual sets of counting equipment will lead to sources of error. The usual procedure is to divide the foils into groups, each group being counted with one set of equipment. A master foil or uranium standard giving approximately the same count rate as the foils themselves is circulated through each set of equipment so that corrections can then be made for individual differences in the equipment. In addition a sub-master foil is counted several times in one set of equipment so that errors due to changes in stability may be corrected. The following table is a typical counting routine assuming 32 foils and four sets of equipment. The foils are divided into groups of eight. The foil A1 of the first group is circulated through all four sets of equipment and foils A2, B1, C1 and D1 are counted several times during the measurements.

TABLE 6.4. Foil Counting Routine for 32 Foils with 4 Sets of Counting Equipment

Set 1	Set 2	Set 3	Set 4
A1	—	—	—
A2	A1	—	—
A3	B1	A1	—
A4	B2	C1	A1
A2	B3	C2	D1
A5	B4	C3	D2
A1	B1	C4	D3
A6	A1	C1	D4
A7	B5	C5	D1
A8	B6	A1	D5
A2	B7	C6	A1
—	B8	C7	D6
—	B1	C8	D7
—	—	C1	D8
—	—	—	D1

A number of correction factors must be applied to the activities obtained by the above method of counting. These are due to background, counter dead-time, statistics and decay.

(a) *Background corrections*

The counting background is reduced to a minimum by using lead castles to shield the counter from external influences. However, a background count always remains, especially when using standard geiger counters. In order to correct for this effect, the background must be measured and subtracted from the measured count rate. If the count rate on the foil is very high (say 3000 counts/min) then the background correction will be negligible and may be neglected.

(b) *Counter dead-time correction*

On account of the electronic system of counting the pulses, it is necessary that the equipment be rendered inoperative after each pulse from a geiger counter. This dead-time effect is produced by a probe unit and enables the equipment to recover itself in time to count the next pulse. Any pulse arriving during the dead-time will obviously not be counted. Hence the measured count rate must be corrected for the number of pulses which were missed during the dead-time.

The true count (corrected count) is given by

$$N_t = \frac{N_0}{(1 - N_0 t)} \qquad (6.11)$$

where t is the dead-time (set by the electronic equipment),
N_0 the observed count rate.

(c) *Statistical error*

The production of β-particles in the activated foils is of a random nature. The error is taken as $\pm \sqrt{N}$ where N is the measured count rate. A total of 10,000 counts will be necessary to achieve a 1% statistical accuracy.

(d) *Decay corrections*

If large numbers of foils are being counted and the total counting period is significant in comparison with the half-life of the foil material, then corrections for the decay of the foil activity during the counting must be applied. It is usual to note the time at which the foil counting commences and then apply corrections using equation (6.10) in which T_1 and T_2 are constant for each foil and T_3 varies.

In order to measure the neutron flux distribution in a reactor, a whole series of foils, all carefully numbered, are distributed in the core. The foils are then irradiated by either balancing the reactor at steady power level (i.e. constant neutron flux) for a sufficient time to irradiate the foils or by allowing the neutron flux to increase with a stable period (constant core reactivity) up to a sufficiently high power level.

The magnitude of the steady state power level and the peak power level attained during the transient method are dependent upon the following factors:

1. The nuclear properties of the foil used (i.e. half-life, absorption cross-section, etc.).
2. The time taken following the shut-down to remove the foils and transfer to the counting laboratory.

Regarding the first factor less irradiation is required for foils with large absorption cross-sections and short half-lives. However, the time taken to remove foils from the core can be quite considerable. In all cases access to the core will be required and in the case of a large power reactor it is often difficult and time consuming to remove the foils from their locations within the core. Withdrawal times can vary between 1 to 2 hr for critical assemblies to as much as 12 to 16 hr for large reactors. Consequently the irradiation flux level and the total irradiation time is usually determined by the total time for foil withdrawal from the reactor.

(e) *Irradiation at constant power level*

In this case, it is essential to raise the reactor power to the steady level as quickly as possible and to shut down after the irradiation as quickly as possible. If this can be achieved then the amount of induced activity of the foil due to raising and lowering the power is negligible compared to the activation at the steady level. Hence T_1 is measured from the moment the steady power level is attained to the moment of shut down. Using equation (6.10) a value for ϕ may be determined from the measured foil activity.

(f) *Irradiation by the transient method*

In this case the power is allowed to double up to a predetermined peak power level and then the reactor is shut down.

This method is particularly useful for flux measurements in cores with a total excess reactivity such that a doubling time of not less than 10 to 15 sec is achieved with all controlling absorbers fully withdrawn. In this case flux distortions due to the effect of the control rods do not occur. Distortions of this nature must occur in the steady state method and lead to errors in deducing the power level of a reactor from flux measurements.

Let us assume that at time t_0, the power is equivalent to a flux of ϕ_0.

Then at time t, corresponding to n doubles of the flux:

$$\phi_t = 2^n \phi_0$$

After time T, corresponding to N doubles:

$$\phi_{\text{peak}} = 2^N \phi_0$$

Hence

$$\phi_t = \frac{2^n}{2^N} \phi_{\text{peak}}$$

If t_D is the reactor doubling time, then

$$t = n t_D$$

and

$$T = N t_D$$

The average flux during the irradiation

$$= \frac{\int_0^T \phi \, dt}{\int_0^T dt}$$

i.e.
$$\phi_{\mathrm{av}} = \frac{\phi_{\max}}{N \log_e 2} \left(1 - \frac{1}{2^N}\right) \qquad (6.12)$$

For N very large

$$\phi = \frac{\phi_{\max}}{N \log_e 2} \qquad (6.13)$$

During the irradiation the number of doubles N is determined and ϕ_{av} is obtained using equation (6.10) and the induced activity on the foils. Hence $\phi_{\max}$ may be deduced.

(g) *Cadmium ratios*

By irradiating foils of a particular material, such as manganese, in a reactor then the foil activities will be a measure of the total neutron flux at the position of the foils. The manganese foil is activitated by thermal and epi-thermal neutrons and a resultant total flux is therefore determined. If, for instance, the flux distribution measurements are to be used for calculating the total power level of the core then a measure of the thermal neutron flux distribution is required. This can be obtained by carrying out two irradiations with foils placed in identical positions first bare and then covered completely with a layer of cadmium.

It has been shown experimentally that a layer of cadmium of thickness 0·025 in. to 0·030 in. will completely absorb all thermal neutrons in a given flux. Hence the induced activity measured on the foils irradiated under cadmium is a measure of the epi-thermal neutron flux or epi-cadmium flux as it is sometimes called. Using this value and the value of the total flux obtained from the bare foil measurements, a thermal flux may easily be deduced. It is usual during under-cadmium irradiations to place at least one bare foil in the core in an identical position to a foil during the bare irradiation in order to correct for any changes in absolute flux magnitude during the measurements.

The ratio of the thermal flux to the epi-thermal flux is known as the cadmium ratio.

It should be noted that in regions where there is a large cadmium ratio a much greater flux or irradiation time will be required to obtain a reasonable induced activity on the under-cadmium foils. Some relaxation of accuracy is obviously permitted however if the epi-thermal flux provides only a small contribution to the total flux.

6.3.5. Determination of the Reactor Power Level

The reactor power level may be determined from flux distribution measurements across the core in the horizontal and vertical planes.[6, 7, 10] Foils are placed in position and irradiated either by the steady power level method or, preferably, by the transient method. Two irradiations are necessary, one with bare and one with the foils under cadmium. From the resultant thermal neutron flux distributions the reactor power may be calculated by integrating under the flux distributions and multiplying by the thermal neutron fission cross-section for the core materials.

It should be realised that this method is not accurate and accuracies of the order of 10 to 15% are the best that can be achieved. In the case of power reactors and research reactors in which a temperature rise across the core is produced and measured then a heat balance method (see Chapter 9) is the most accurate method of determining the power output of the core. However, in the case of small zero energy reactors or research reactors in which temperature instrumentation is inadequate a nuclear method is employed.

In the case of a large power reactor, it is usual to determine the power level by foil techniques, as measured by the temporary instruments placed in or near the "just critical" core. In carrying out the reactor physics measurements and other nuclear commissioning tests, the reactor is allowed to double up to a certain pre-determined power level many hundreds of times. The pre-determined power level is of course only an estimate and it is essential at the first convenient stage in loading above criticality to determine the power level. Hence, after measuring the power level the doubles may be terminated at such a level that excessive irradiation of the fuel is prevented during the commissioning tests.

The power level is often determined by comparison with the results obtained on another similar reactor. For instance, if an operating reactor already exists with the same core and fuel arrangement then power calibration is a simple and easy experiment to perform. An absolutely calibrated foil (gold is usually used for absolute measurements as its cross-section is known accurately) is irradiated in the operating reactor at a particular power level and particular control rod pattern for a specified time. The activity on the foil is determined. The same foil is then irradiated in the reactor, which is being commissioned, with as near as possible the same core conditions for the same time. The activity on the foil is determined and a power calibration obtained by comparison of the two activities.

6.3.6. Programme for Reactor Physics Experiments

As discussed in the preceding sections, large numbers of experiments are carried out after the critical size has been determined. In order to make most use of the time available for these tests and so that the fuel loading does not have to be reduced at any stage, careful planning of the physics tests is of the utmost importance. In order to give an example of a fairly complex programme the procedure for a large power reactor is given in Table 6.5. The fuel loading stages are referred to as numbers of channels where in this case the amount of fuel per channel would be of the order of 80 kg of natural uranium.

From the doubling time measurements the variation of k_{eff} with β^2 may be determined, if the extrapolated radius is known for each stage. The extrapolated radius is equal to the physical radius plus the extrapolation length λ. The extrapolation length for the core is determined by extrapolating the measured flux distributions in the horizontal plane.

6.3.7. The Measurement of Reactor Lattice Constants

In many cases it is possible to determine values of the Laplacians of the core, the value of the infinite multiplication factor and values for the migration areas.[15] These experimental values, although not necessary from an operating point view, are of interest in checking the design data. In order to achieve reasonable experimental accuracy it is necessary to be able to load fuel in a uniform manner and in quantities such that the excess reactivity of the core may be varied in small amounts. Obviously the full treatment is therefore only possible for large cores and then only if an efficient method of poisoning the core uniformly is available.

For a large cylindrical reactor the expression for the effective multiplication factor is given by equation (6.2)

i.e.
$$k_{eff} = \frac{k_\infty}{1 + \alpha^2 M_z^2 + \beta^2 M_r^2}$$

TABLE 6.5. TYPICAL REACTOR PHYSICS PROGRAMME

Stage	Channels loaded	Channels loaded above criticality	Excess reactivity (%)	Doubling time (sec)	Experiments planned
1	740	0	0	∞	Completion stage of loading to critical size. Estimated critical size—744 channels
2	754	10	0·071	100	(a) Doubling time measurement. (b) Air pressure coefficient of reactivity at 3 different pressures below atmospheric (c) Power calibration by foil techniques (d) Flux distribution near and reactivity effect of an empty channel
3	758	14	0·085	68	Doubling time measurement
4	762	18	0·099	53	Doubling time measurement
5	766	22	0·118	41	Doubling time measurement
6	770	26	0·132	32	(a) Doubling time measurement (b) Flux distribution in horizontal and vertical planes (c) Flux distribution about and reactivity effect of partially inserted control rod
7	778	34	0·163	18	Doubling time measurement
8	788	44	0·205	13	Doubling time measurement

By differentiating with respect to β^2, we get

$$\frac{dk_{eff}}{d(\beta^2)} = -\frac{M_r^2 k_\infty}{(1+\alpha^2 M_z^2+\beta^2 M_r^2)^2} \qquad (6.14)$$

and at criticality

$$k_\infty = 1+\alpha_c^2 M_z^2+\beta_c^2 M_r^2$$

or

$$\left.\frac{dk_{eff}}{d(\beta^2)}\right|_c = -\frac{M_r^2}{k_\infty}$$

or

$$M_r^2 = -k_\infty \left.\frac{dk_{eff}}{d(\beta^2)}\right|_c \qquad (6.15)$$

Thus by varying the loading above criticality at constant height and measuring the doubling time it is possible to determine k_{eff} and β^2 at each stage of loading.

Hence, if k_{eff} is plotted against β^2 near to criticality (i.e. $k_{eff} \doteqdot 1\cdot0030$), the slope of the graph at the critical point (S_r) may be determined, and so

$$M_r^2 = -k_\infty S_r \qquad (6.16)$$

If the height of the core is varied at constant radius and doubling times measured at several stages, then similarly we get

$$\left.\frac{dk_{eff}}{d(\alpha^2)}\right|_c = -\frac{M_z^2}{k_\infty} \qquad (6.17)$$

If k_{eff} is plotted against α^2 and the slope of the graph at the critical point (S_h) determined, then

$$M_z^2 = -k_\infty S_h \qquad (6.18)$$

By substitution, we get

$$k_\infty\left[\frac{1}{k_e}+\alpha^2 S_h+\beta^2 S_r\right] = 1 \qquad (6.19)$$

Hence, it should be possible to measure values of α^2, β^2, M_r^2, M_z^2 and k_∞ by using equations (6.16), (6.18) and (6.19).

Let us consider as an example a gas cooled graphite moderated natural uranium reactor in which each channel contains twelve fuel elements. The first stage in the experiment would be to carry out an approach to criticality loading only 10 elements into each channel. The critical size of the core would be determined and then the variation of the k_{eff} of the core with fuel loaded in the supercritical region would be carried out. Fuel would be loaded in stages with 10 elements per channel and doubling time measurements carried out at each stage. In addition flux distribution measurements would be made in both the vertical and horizontal planes and from these the radial and axial extrapolation lengths determined. Hence values of β^2 can be obtained at each loading stage. The variation of k_{eff} with β^2 will be plotted and the slope of the curve at criticality determined (i.e. S_r in equation (6.16)).

On completion of these measurements the height of the core will be varied at constant radius. This is carried out in two stages by adding an additional fuel element to each channel in each stage, hence loading 11 and 12 elements per channel at a fixed core radius. It is necessary to be able to measure k_{eff} at these axial core loadings.

With 10 elements per channel loaded a doubling time will have been measured at the end of the radial loading. Hence, the excess reactivity of the core at a particular extrapolated radius will have been measured. If this corresponds to a doubling time of 20 sec, then k_{eff} will be approximately 1·0016. If an additional element per channel (i.e. 11 elements per channel) is loaded over the radius, then the excess reactivity of this core will be far too large to enable a doubling time to be measured with all the control rods fully withdrawn. The distributed poisoning technique is used to take up the excess reactivity and so enable a doubling time to be measured.

During the measurements with 10 elements per channel

an air pressure coefficient of reactivity will be determined. Using this value and the estimated value of k_{eff} with 11 elements per channel, then the pressure required to absorb the excess reactivity may be calculated. It is usual to absorb all excess reactivity except approximately 0·1%; so that with the control rods withdrawn a doubling time of approximately 30 sec is achieved. Then if α_p is the measured pressure coefficient of reactivity, P the balancing pressure and δk_e the excess reactivity corresponding to the measured doubling time; then

$$k_{\text{eff}} = \alpha_p P + \delta k_e \qquad (6.20)$$

The experiment is then repeated with 12 elements loaded per channel and k_{eff} again determined. From these results the variation of k_{eff} with α^2 may be plotted and the slope of the curve (S_h) at criticality determined. By substitution in equations (6.16), (6.18) and (6.19) values for α^2, β^2, M_r^2, M_z^2 and k_∞ may be evaluated.

These experiments may also be carried out with other types of reactors using different techniques for varying the radius and height of the core.

In the case of a water moderated enriched uranium reactor with single elements per lattice position, the variation of effective core height may be achieved by varying the physical height of the water in the core itself. Distributed poisoning would be achieved by using strips of absorber (silver, gold or borated plastic) distributed uniformly in the core. An absorber coefficient of reactivity (i.e. reactivity change per unit of absorber) may be determined in a similar way to the air pressure coefficient method, by measuring the change in doubling time and hence reactivity produced by a small change in the loaded absorber.

In most reactors the value of S_r is obtained from the determination of the variation of k_{eff} with core loading above criticality. The theoretical value of the ratio M_r^2/M_z^2 for graphite moderated natural uranium reactors is known to a high degree of accuracy. This value has been confirmed in a large variety of sub-critical and exponential lattices. Hence by using the theoretical value of M_r^2/M_z^2 in conjunction with the above equations, and the measured value of S_r, values of k_∞, M_r^2 and M_z^2 may be obtained.

It is more usual to make a compromise in this way to determine the core parameters and usually only the variation of k_{eff} with core loading is carried out experimentally.

6.4. THE ESTIMATION OF k_{eff} FOR THE FULLY LOADED REACTOR FROM "JUST CRITICAL" MEASUREMENTS

One of the objects of loading a new reactor core to criticality, determining the critical size and measuring the variation of k_{eff} with loading is to be able to estimate at an early stage the total excess reactivity of the fully-loaded reactor. Hence, loading the reactor to full size may be carried out with a knowledge of the total excess reactivity and so an unforeseen dangerous situation will not arise.

In the case of water moderated enriched uranium reactors, an excess reactivity of 2·0% may be achieved with a comparatively small increase in mass above the critical core. As an example an excess reactivity of 2·0% could be attained by loading approximately 300 g of U-235 above criticality for the Queen Mary College critical assembly.

Hence, in this type of core the total excess reactivity of the fully loaded reactor may be estimated with a high degree of accuracy from the mass coefficient of reactivity measured at the "just critical" stage.

In the case of a large power reactor the problem is more

difficult. Let us consider the case for a graphite moderated natural uranium reactor in which the central region contains flattening material. Flattening can be achieved by inserting absorbers uniformly in the central region. Additional excess reactivity must then be built into the core to compensate for this absorber. If therefore the central region of the core contains absorbing material up to a radius, R_F, the core is divided into two radial regions, each having different radial parameters.

Let region 1 represent the central region containing absorber and region 2 the outer non-flattened region.

The measurement of critical size and the variation of k_{eff} with fuel loaded will have been determined in region 1. Hence, if the slope S_1 of the k_{eff} against β_1^2 is measured in this region, we have that

$$M_r^2 = -k_{\infty 1}S_1$$

Assuming that the ratio M_z^2/M_r^2 is known and equal to x, then

$$M_z^2 = -k_{\infty 1}S_1 x$$

If α_c^2 and β_c^2 are the Laplacians for the reactor at criticality, then

$$k_{eff\,1} = \frac{k_{\infty 1}}{1 + \alpha_c^2 M_z^2 + \beta_c^2 M_r^2} \tag{6.21}$$

and $k_{\infty 1}$ may be determined.

During the "just critical" measurements with region 1 materials, the reactivity worth of interchanging a channel typical of the outer region (without absorber) will be determined by a difference in doubling time method. Hence the reactivity effect of completely changing to a critical core consisting of region 2 type fuel may be calculated by weighting the measured reactivity effect for one channel with respect to the square of the radial flux distribution over the "just critical" core. Let this total reactivity change be Δk_e.

Now let $k_{\mathrm{eff}\,2}$ be the effective multiplication factor for the type 2 fuel core.

Then

$$k_{\mathrm{eff}\,2} = k_{\mathrm{eff}\,1} + \Delta k_e \qquad (6.22)$$

and hence, by substituting in equation (6.21),

$$k_{\mathrm{eff}\,2} = \frac{k_{\infty 2}}{1 + \alpha_c^2 M_z^2 + \beta_c^2 M_r^2} \qquad (6.23)$$

and $k_{\infty 2}$ may be deduced.

Assume that the fully-loaded radius of the core is R_L and that it contains type 1 fuel up to radius R_F and type 2 fuel from radius R_F to R_L. The value of β_L^2 for the fully-loaded reactor may be deduced assuming an extrapolation length and a flux distribution for the fully loaded core consisting of one type of fuel. Then, if the core contains only type 1 fuel:

$$k_{\mathrm{eff}\,1}(\text{fully loaded}) = \frac{k_{\infty 1}}{1 + \alpha_c^2 M_z^2 + \beta_L^2 M_r^2} \qquad (6.24)$$

and if only type 2 fuel is used:

$$k_{\mathrm{eff}\,2}(\text{fully loaded}) = \frac{k_{\infty 2}}{1 + \alpha_c^2 M_z^2 + \beta_L^2 M_r^2} \qquad (6.25)$$

The effect of interchanging the complete core from region 1 type to region 2 type is given by

$$k_{\mathrm{eff}\,2}(\text{fully loaded}) - k_{\mathrm{eff}\,1}(\text{fully loaded}) = \Delta K_e \quad (6.26)$$

So, the effect of only interchanging to radius R_F is

$$\Delta K_e \frac{\displaystyle\int_0^{R_F} \phi^2 r \, dr}{\displaystyle\int_0^{R_L} \phi^2 r \, dr} = I \qquad (6.27)$$

Then, the total excess reactivity of the fully loaded reactor is given by

$$k_{\text{eff}}(\text{reactor}) = k_{\text{eff}\,2}(\text{fully loaded}) - I \qquad (6.28)$$

6.5. LOADING TO FULL SIZE

The loading of a reactor to its full size such that all available fuel element positions are completely filled with fuel is only carried out after a full investigation and analysis of the "just critical" core measurements. It is important to remember that a reactor should not be fully loaded until:

(1) An accurate estimate of the total excess reactivity has been obtained.

(2) An initial estimate of the total worth of the control rod system has been made.

From a safety point of view both of the above factors must be known so that there is no possibility whatever of unintentional criticality being achieved during the loading, or that the permitted excess reactivity limit is exceeded. In the preceding section the estimation of the excess reactivity has been considered in detail.

Early estimates of the worth of a partially inserted or fully inserted rod or rods may be made during the approach to criticality and these estimates are used to deduce the total shut-down capacity. However, this procedure cannot be carried out with any degree of accuracy and therefore it is usual to proceed with care when loading to full size.

For a gas-cooled, graphite moderated natural uranium reactor a large amount of fuel is required to be loaded above the critical size to achieve the operational amount of excess reactivity. For instance the critical size is of the order of 50 tonnes of natural uranium, whereas a total of 250 to 300 tonnes will be required to achieve an excess reactivity of the order of 5·0%.

The loading to full size is carried out in stages of approximately 300 channels (i.e. 25 tonnes of natural uranium). At the end of each stage geometrical symmetry should be maintained and neutron flux measurements are then made with all the control rods inserted and with various combinations and groups of rods withdrawn. Approach to criticality curves are plotted in a similar way to the initial approach to criticality. In this way by extrapolating the approach curves, the amount of fuel to be loaded to achieve criticality for each combination of inserted control rods and for all the control rods inserted may be predicted. This method ensures absolute safety during the loading of the core to full size.

With a smaller enriched uranium water reactor the loading to full size will be carried out in two or three stages depending upon the design of the individual fuel elements. The final stage will be such that the estimated total excess reactivity does not exceed the permitted limit. The excess reactivity must be determined experimentally immediately the core has been loaded. This is done by carrying out an approach to criticality on rod withdrawal (i.e. measure the subcritical neutron flux level near to the core as the rods are withdrawn in discrete stages). The withdrawal position of the control rods for criticality will therefore be predicted and hence an estimate made of the excess reactivity. This would be confirmed subsequently on completion of the calibration of the control rod system.

REFERENCES

1. Manual for the operation of research reactors. *I.A.E.A. Technical Reports Series* No. 37 (1965).
2. KEEPIN, G. R. *Physics of Nuclear Kinetics*, p. 189. Addison-Wesley (1965).

3. KING, L. D. P. *Proc. of First Geneva Conf.* **2**, 372. U.N. (1956).
4. PAXTON, H. C. and KEEPIN, G. R. Criticality. U.S.A.E.C. Project SIFTOR. M.I.T. Press (1964).
5. GLASSTONE, S. and SESONSKE, A. *Nuclear Reactor Engineering.* Van Nostrand (1964).
6. STURM, W. J. (Ed). Reactor laboratory experiments. *Argonne Nat. Lab. Report* ANL-6410 (1961).
7. HOAG, J. B. *Nuclear Reactor Experiments.* Van Nostrand (1958).
8. BLAND, A., KENNEDY, D. J. and MENARRY, A. Physics aspects of Berkeley and Bradwell Commissions. Parts I and II. *Conf. on Physics of Graphite Moderated Reactors.* Inst. of Physics. April (1962).
9. ANDERSON, H. L. *et al. Phys. Rev.* **72**, 16 (1947).
10. HUGHES, D. J. *Pile Neutron Research.* Addison-Wesley (1953).
11. THIE, J. A. Theoretical reactor statics on kinetics of boiling reactors. Second Int. Conf. Geneva. Paper 638.
12. DE SHONG, J. A. Styrofoam simulation of boiling and temperature effects in the E.B.W.R. *Argonne Nat. Lab. Report* ANL-6229 (1960).
13. UTHE, P. M. Attainment of neutron flux spectra from foil activations. *U.S.A.F.I.T T. echnical Report* 57–3.
14. WEINBERG, A. M. and WIGNER, E. P. *The Physical Theory of Neutron Chain Reactors.* Univ. of Chicago Press (1958).
15. *Transactions of American Nuclear Society* 1960/1966. Sections (a) Critical experiments; (b) Nuclear parameter measurements; (c) Neutron flux measurements; (d) Experimental methods in reactor physics.

CHAPTER 7

Control Rod Calibrations and Temperature Coefficient Measurements

REACTIVITY measurements carried out on reactors operating at power are determined by the relative movements of the various absorbers in the reactor control system.[1, 2, 3, 4, 14] It is important for the operating staff to have accurate information about reactivity changes due to xenon poisoning, long-term reactivity effects, movement of absorbers, movement of flattening material and movement of control rods individually or in groups. It is therefore essential to obtain as much detailed experimental information about the reactivity worths of control rods as possible.

The reactivity worth of the control rods should be determined for individual rods and for the groups and combinations of rods which are likely to be used in normal operation. The total worth of all the rods will enable an estimate to be made of the total shut-down capacity of the core. This is defined as the reactivity safety margin of the the reactor core at shut-down and is the reactivity difference between the total worth of the control system and the total excess reactivity of the core. It is usual to design the control system such that for the cold unpoisoned reactor this margin is at least 1·5% in reactivity.

It is more convenient to carry out control rod calibrations before the reactor has been brought to power. Most of the methods involve measurements requiring access to the core. The problems of control rod calibrations for a large reactor with a complex control system are very numerous. The main problem with a flexible reactor design is to decide in advance which will be the most likely operational arrangements of the rods. This is not so easy as appears at first sight. Many research reactors also have a flexible core design and cores are built in a variety of ways around a control rod system. Obviously the worth of individual rods and groups of rods in such a system is very dependent upon the geometrical shape of the core and its position relative to the control rods.

In the case of a large natural uranium power reactor, control rods are usually distributed uniformly over the core lattice and therefore there may be as many as 130 rods or more. During operation most of these rods will be fully withdrawn, but groups of rods are used to control the reactor and to flatten the flux distribution. The number and distribution of rods in these groups varies during operation and it is difficult to predict these variations in advance.

In addition for all types of reactors the number and distribution of the control rods will vary throughout the life of the reactor as the long-term reactivity effects change the overall excess reactivity of the core.

One answer to this problem is of course to calibrate all possible combinations of rods. However, this is far too time-consuming. A compromise is usually adopted and it is possible by calibration of carefully chosen rod combinations to be able to extrapolate during operation to another system. A check of the calibrations may be carried out by calibrating the rods at power. However, as will be seen later (Chapter 9) these methods are limited in application.

A large number of different methods are available for calibrating control rods. They have a variety of uses and the method to be adopted is very dependent upon the type of reactor design and the total reactivity worth of the individual rods.

The most widely used methods are:

1. Doubling time method.
2. Intercalibration method.
3. Rod drop technique.
4. Source jerk method.
5. Rod oscillator method.
6. Pulsed neutron source technique.
7. Distributed poison technique.

These will be dealt with in turn.

7.1. DOUBLING TIME METHOD

The method of rod calibration by measuring reactor doubling times associated with an excess reactivity is limited in scope and application. It is not safe or feasible to attempt to measure doubling times less than 10 to 15 sec and consequently the maximum possible reactivity worth measurable by this method is approximately 0.20%. It is useful for measuring the total worth of small absorbers or fine control rods used in many research reactors. The partial worth of control rods or combination of control rods over a small range of travel near the criticality point may also be measured by this method.

Prior to commencing a calibration by this method the inhour relationship (equation (6.7)) must be derived for the core.[1]

Let us consider first the calibration of a rod of total worth approximately 0.20%. The technique is to obtain a power balance for the reactor by rod withdrawal leaving the rod to be calibrated fully inserted. On completion of this measurement the reactor is shut down and the rod to be calibrated is withdrawn by say 10% of its total distance of travel. The rod is then held in this position and the other control rods withdrawn to the previously measured balance point. The reactor flux level will then rise with a stable period corresponding to the excess reactivity released in moving the experimental rod by 10%. The doubling time is measured up to a pre-determined power level and then the reactor is shut down. From the inhour relationship the reactivity worth of the first 10% of the experimental rod may be determined.

The experiment is repeated withdrawing the experimental rod in stages of 10% and measuring the doubling time at each stage until the rod is fully withdrawn. Hence in this way a complete calibration curve for the rod is measured.

The experiment must be carried out with care. Before proceeding to the next 10% withdrawal, it should be verified from the commenced control rod calibration curve that a reasonable doubling time should be obtained at the next stage. On no account should any measurement be attempted if there is likely to be a short doubling time giving rise to a reactor trip.

The doubling time technique may be used to calibrate a control rod over a part of its travel.[3] For instance, having determined a reactor balance point, a rod may be moved a small amount beyond the balance point with the reactor shut down with other rods. The shut-down rods are withdrawn and a doubling time measured. The experimental rod may be moved again and so on and the procedure repeated until a doubling time of 10 to 15 sec has been reached.

The main sources of error in this type of measurement, apart from the normally expected errors involved in measuring time, are due to errors involved in calculating the reactivity–period relationship using the inhour equation. Errors are due to the fact that the inhour equation does not allow for certain transient terms in the equations for the reactivity-period relationship. Hence the calculations are only accurate provided that the initial transients have died away. These errors may be avoided by only including results for which successively measured doubles are within the experimental accuracy.

A second source of error may arise if a set of average delayed neutron periods are used to calculate the effect of the delayed neutrons in the inhour relationship. If precise reactivity measurements (better than $\pm 1\%$ accuracy) are required then the correct delayed neutron parameters for each individual precursor must be used.[1]

7.2. INTERCALIBRATION METHOD

As its name implies this method involves the calibration of a control rod or group of rods relative to a rod or group of rods which has already been calibrated previously by an absolute method. The intercalibration method is a relative one and hence some other form of calibration must also be carried out if absolute values are required.

The procedure usually adopted is to obtain a reactor balance point at a low power level using the rod or rods already calibrated with the rod to be calibrated fully inserted. The calibrated rods are then inserted by a given amount (10% of their total distance of travel) and the low power balance position maintained by withdrawing the uncalibrated rod. If for a movement of 10% of the calibrated rods, a balance is maintained by a movement of 20% of the uncalibrated rod, then the reactivity worth of the first 20% of the uncalibrated rods is equivalent to a 10% movement of the calibrated rods.

The procedure may be repeated over the complete withdrawal distance for the uncalibrated rods.

This method is not accurate on account of the fact that during the calibration the flux distribution across the core will be radically different for the various balance positions. In addition the reactivity shielding effects in groups of rods cannot be allowed for in this method (i.e. the total reactivity worth of a group of rods is less than the sum of the individual rod worths).

However, the method is an extremely useful one for obtaining an initial indication of control rod worths.

7.3. ROD DROP TECHNIQUE

The rod drop method for measuring negative reactivity step changes in a reactor involves observing the decay of the reactor power following the tripping of a rod or groups of rods.[1, 7] The method is one which may be used at any time during the operational life of the reactor and has the advantage that it is both fairly easy and quick to perform. It is also widely used to determine the total shut-down capacity of a control rod system provided this is not greater than 4 to 5% in reactivity.

The equations describing the kinetic behaviour of a reactor are given by Keepin[1] and are

$$\frac{dN}{dt} = k \frac{(1 - \bar{\gamma}\beta) - 1}{l} N + \sum_i \lambda_i C_i + S \qquad (7.1)$$

and

$$\frac{dC_i}{dt} = \frac{\bar{\gamma}\beta k}{l} N - \lambda_i C_i \qquad (7.2)$$

where C_i is the precursor concentration of the ith delayed neutron group,

λ_i the precursor decay constant of the ith group,

S the neutron source contribution,

$\bar{\gamma}$ the average effectiveness in producing fission of delayed neutrons with respect to prompt neutrons,

β the delayed neutron fraction compared with all fission neutrons,

$\bar{\gamma}\beta$ the effective delayed neutron fraction,

N the neutron density (time dependent quantity)

l the prompt neutron lifetime, and

k the neutron multiplication factor.

Consider a reactor operating at a steady power level (P_0) with a corresponding neutron density (N_0) with $N_0 > S$ which is shut down rapidly by the sudden insertion of a neutron absorber. The introduction of the absorber will produce a negative reactivity change which will cause a step change in the neutron density. Within a few prompt neutron lifetimes after the step change the neutron density will fall to a quasi-equilibrium level which will then be decreased by delayed neutron decay.

At the equilibrium level prior to the step change the kinetic equations (7.1) and (7.2) become

$$\frac{dN}{dt} = 0 = \frac{(k_{p_0} - 1)}{l} N_0 + \sum_i \lambda_i C_i \qquad (7.3)$$

and

$$\frac{dC_i}{dt} = 0 = \frac{\bar{\gamma}\beta}{l} N_0 - \lambda_i C_i \qquad (7.4)$$

where $k = 1$ and k_{p_0} is the value of the prompt neutron multiplication factor. At equilibrium the contribution of the prompt neutrons is represented by k_{p_0} which is given by $k_{p_0} = k(1 - \bar{\gamma}\beta)$.

Thus, from (7.3)

$$N_0 = \frac{l \sum_i \lambda_i C_i}{1 - k_{p_0}} \qquad (7.5)$$

Immediately following the step change the neutron density falls to N_1 and if it is assumed that at this stage C_i is unchanged then

$$N_1 = \frac{l \sum_i \lambda_i C_i}{1 - k_{p_1}} \qquad (7.6)$$

where k_{p_1} is the value of the prompt neutron multiplication factor at the new level and $k_{p_1} < k_{p_0}$ and hence

$$\frac{N_0}{N_1} = \frac{1 - k_{p_1}}{1 - k_{p_0}} = \frac{1 - k(1 - \bar{\gamma}\beta)}{\bar{\gamma}\beta} \simeq 1 + \frac{\Delta k}{k\bar{\gamma}\beta} \qquad (7.7)$$

and so the reactivity change in percentage is given as

$$\frac{\Delta k}{k} = \left(\frac{N_0}{N_1} - 1\right)\bar{\gamma}\beta \times 10^2 \qquad (7.8)$$

Hence by determining N_0 and N_1 the value of the step change in reactivity may be obtained. The value of N_1 is determined by plotting the value of the power level against time following the step change in reactivity and extrapolating the curve back to determine the value of N for $t=0$, the commencement of the drop.

The chief source of error in this method is due to the difficulty in extrapolating the decay curve to determine the step change in reactivity from the delayed neutron decay characteristics. The ratio of the neutron density at time t after the step change in reactivity (N_t) to the initial density (N_0) may be determined using the delayed neutron data in the appropriate kinetic equations. A typical curve relating the ratio of the power at times of 15, 30, 50, 100 and 200 sec to the initial power level with negative reactivity is given in Fig. 7.1.

The reactivity worth of a control rod or group of rods may therefore be determined by balancing the reactor at a given

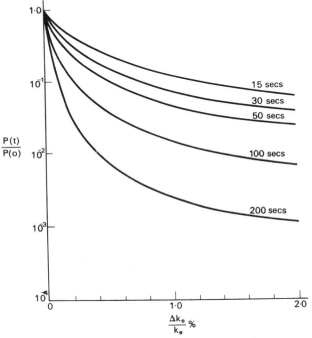

FIG. 7.1. Typical power ratio–negative reactivity decay curve

power level with the rods to be calibrated fully withdrawn. The rods are then tripped into the core and the resultant power decay curve plotted. From the ratio N_t/N_0 at various times after the trip the reactivity change produced and hence the worth of the rods may be determined. A calibration may be carried out by repeating the drops with the rods withdrawn by varying amounts provided that a balance at power can be achieved.

7.4. SOURCE JERK METHOD

The source jerk method of calibration is similar to the rod drop method in that it makes use of the reactor decay curve. In this case however the measurements are carried out with the reactor in a subcritical state and therefore has the advantage that results can be obtained over a large range of rod movement.

The technique is to introduce a neutron source into the reactor in the subcritical state and obtain measurements of the neutron density in and near the core. The source is then withdrawn rapidly (jerked) from the core and the variation of neutron density following the jerk is measured.

If the source density is S neutrons per cm³ and the neutron density with the source inserted n_0, then from the kinetic equations (equations (7.1) and (7.2)) we get

$$\frac{dn_0}{dt} = 0 = \frac{k_{p_0}-1}{l} n_0 + \sum_i \lambda_i C_i + S \qquad (7.9)$$

and

$$\frac{dC_i}{dt} = 0 = \frac{\bar{\gamma}\beta k_0 n_0}{l} - \lambda_i C_i \qquad (7.10)$$

where k_0 is the multiplication constant for the core.

Hence:

$$n_0 = \frac{l \sum_i \lambda_i C_i + lS}{1 - k_{p_0}} \qquad (7.11)$$

Immediately the source has been removed the neutron density will fall to a new level n_1 given by

$$n_1 = \frac{l \sum_i \lambda_i C_i}{1 - k_{p_0}} \qquad (7.12)$$

and hence:

$$\frac{n_0}{n_1} = 1 + \frac{S}{\sum\limits_{i} \lambda_i C_i} \tag{7.13}$$

and using equations (7.9) and (7.10) we get

$$\frac{n_0}{n_1} = 1 + \frac{1 - k_0}{k_0 \bar{\gamma} \beta}$$

or

$$\frac{\Delta k}{k} \% = \left(\frac{n_0}{n_1} - 1\right) \bar{\gamma} \beta \times 10^2. \tag{7.14}$$

Thus by determining n_1 and n_0 in or near the core before and after the source is removed then the negative reactivity of the core can be determined.

The above expressions assume that the source is removed instantaneously and therefore for accurate measurements the source must be withdrawn as rapidly as possible. Apparatus must be constructed for this purpose. However, it is true to say that small sources can be withdrawn much more rapidly than control rods can drop into the core in the rod drop method. One disadvantage of the method is that flux levels are low and consequently sources of large intensity are required for good results. Adequate results can be achieved with sources emitting 10^5 to 10^6 n/sec and jerk times of the order of 0·02 sec.

7.5. R O D   O S C I L L A T O R   T E C H N I Q U E

If the reactivity of a reactor is varied in a periodic manner as by the periodic motion of a control rod, the reactor power level will also vary periodically. Expressions can be derived from the kinetic equations which describe the behaviour of the power level for a given type of periodic oscillation in reactivity.[1, 8, 9]

If the oscillatory component of the steady-state neutron density is dn at a neutron density level n, then

$$\frac{dn}{n} = W(jw)\, dk \qquad (7.15)$$

where dk is the amplitude of the reactivity oscillation and $W(jw)$ is the reactor transfer function.[1] Expressions can be derived for the reactor transfer function from the reactor kinetic equations for the particular reactor system. Transfer functions may also be determined experimentally from the frequency response to reactivity oscillations of small amplitude. At high frequencies of oscillation, we get

$$dk = \omega l\, \frac{dn}{n} \qquad (7.16)$$

where ω is the angular frequency of oscillation and l is the prompt neutron lifetime. Hence by oscillating a control rod at a known frequency and determining the variation of the power level (dn/n) the reactivity change due to this rod movement may be determined.

The advantages of this method are:

(a) Oscillations can be carried out every few inches over the complete rod travel and hence a complete calibration may be obtained.

(b) The reactor power does not need to be maintained at a steady level during the measurements.

(c) Only a few minutes are required to obtain measurements at each point.

The main disadvantage is that for accuracy the oscillation periods must be small $(\omega \gg \beta/l)$ for the above simple formula to apply and hence elaborate apparatus is required to oscillate the rods to be calibrated.

7.6. PULSED NEUTRON SOURCE TECHNIQUE

In this method a pulse of neutrons is injected into a sub-critical system and the decay of the neutron flux is observed as a function of time after the pulse.[1, 10, 11, 12, 13, 14] Neutron pulses of high intensity are obtained from particle accelerators which utilise the $D(D, n)He^3$ or the $T(D, n)He^4$ nuclear reactions. A number of different types of accelerator are available working on the principle of accelerating deuterium or tritium ions over a voltage in the region of 150 to 200 kV and bombarding a suitable deuterium or tritium target. To utilise this method it must be possible for the target at the end of the accelerator tube to be placed inside the reactor core.

With the control rods inserted into the reactor pulsed neutrons are injected and the neutron flux decay is measured with suitable counters and time analysis equipment. The fundamental mode decay constant for the delayed neutrons α_0 is measured. This is defined as the fractional change in neutron density per unit time. Consequently by plotting the variation of neutron flux or density against time; the decay constant α_0 may be obtained from the slope of the graph. Figure 7.2 indicates a typical measured variation of the decay of the neutrons from a pulse injected into a subcritical system. The neutrons are initially multiplied by the medium (in a period of 3·5 msec) and then decay in a period which is dependent upon the kinetic properties of the reactor system. After 5 msec the rate of decay is reduced due to the effects of delayed neutrons. The slope of the decay curve between points A and B is a measure of α_0.

The relationship between α_0 and the reactivity is obtained from the kinetic equations and is given by

$$\alpha_0 = \frac{1}{n}\frac{dn}{dt} = \frac{(k-1)-k\bar{\gamma}\beta}{l} \qquad (7.17)$$

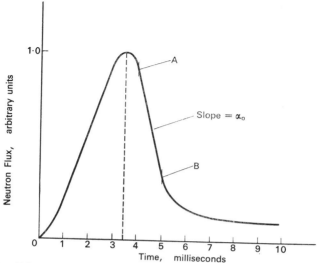

Fig. 7.2. Decay of a pulse of neutrons injected into a subcritical system

The reactivity ϱ in dollars is $\dfrac{k-1}{k\bar{\gamma}\beta}$; and so

$$\alpha_0 = \frac{k\bar{\gamma}\beta}{l}(\varrho-1) \tag{7.18}$$

At delayed critical $\alpha_0 = \alpha_{DC}$ and $\varrho = 0$, and so

$$\alpha_{DC} = -\frac{\bar{\gamma}\beta}{l/k} \tag{7.19}$$

Hence: reactivity ϱ in percentage is given by

$$\varrho = \left(\frac{(\alpha_0 - \alpha_{DC})}{\alpha_{DC}}\right)\bar{\gamma}\beta \times 10^2 \tag{7.20}$$

Thus the reactivity of a system may be obtained by measuring α_0 and α_{DC}. The value of α at delayed critical is deduced by plotting α_0 against rod position and extrapolating to determine the value of α at the previously determined position of the rods at criticality. Hence the rods can be calibrated in the subcritical region.

7.7. DISTRIBUTED POISON METHOD

If a neutron absorber is added uniformly to a reactor core then the control rods will have to be withdrawn to achieve criticality in order to compensate for the reactivity absorbed. By loading a sufficient amount of absorber then it will be possible to balance the reactor with all the control rods fully withdrawn. In this case the total excess reactivity is held by the distributed absorber. The addition of an absorber will change the thermal utilisation factor and the resonance escape probability of the system producing a decrease in the effective multiplication factor. It is possible to calculate the reactivity effect of adding the absorber by determining the change in the thermal utilisation and resonance escape probability from the physical properties of the absorber.[1, 5, 6]

Therefore if absorber is added in stages and the control rods withdrawn to obtain a balance at each stage it is possible to calibrate the control rods over their complete range of travel. The advantage of this method is that it is comparatively easy to calibrate groups and combinations of rods over their complete range. In order to utilise this method access to the core is usually necessary and it is therefore not suitable after the core has been irradiated. In addition the method is only reliable if the absorber or poison can be uniformly distributed throughout the core in such a manner that the neutron flux distribution is unaltered by the addition of the poison. The need for uniform distribution is the main limitation of this method. In water reactors it is usual to use thin strip of silver or stainless steel. In the case of gas-cooled reactors the method has been used extensively by using air at pressure in the vessel. It is found that the absorbing effect of nitrogen in air is sufficient to absorb 4·0% in reactivity with an air pressure of the order of 300 cm of mercury. The advantage of poisoning the core with air is that the poison is added in a very uniform way throughout the reactor.

Although it is possible to calculate the effect of the poison it is more usual to determine a poison coefficient of reactivity which is then used to obtain the control rod calibration curve. The poison coefficient of reactivity is determined by measuring reactor doubling times before and after a small change in the amount of uniformly distributed absorber added to the core. From the difference in doubling time the reactivity effect of changing the absorber may be obtained and hence the reactivity per unit change in absorber calculated.

To illustrate the method let us consider the procedure required for calibrating a group of rods in a gas-cooled reactor. The stages of measurement would be:

1. Withdraw the control rods in the subcritical region with air in the pressure vessel at atmospheric pressure and attain a balance.
2. Withdraw the control rods a small amount beyond the balance point and measure a doubling time.
3. Reduce the air pressure (by exhausting the vessel in this case) in two or three stages and measure a doubling time at each stage.
4. Determine an air pressure coefficient of reactivity from the above results.
5. Pressurise the reactor vessel with air to a certain pressure.
6. Withdraw the control rods to achieve a balance at the new pressure.

7. Repeat the doubling time measurements at the new pressure and again determine an air pressure coefficient of reactivity.

8. Repeat in stages until the control rods are fully withdrawn.

The amount of reactivity absorbed by the air can be determined from the balance pressures and the values of the air pressure coefficient of reactivity. This reactivity is equal to the reactivity released as the control rods are withdrawn.

The total excess reactivity of the reactor is obtained during this calibration from the amount of reactivity absorbed by the air with all the control rods completely withdrawn.

A typical value for the air pressure coefficient is $-12 \cdot 0 \times 10^{-5}$ per cm of mercury pressure and hence $5 \cdot 0\%$ in reactivity could be absorbed by a pressure of just over 400 cm of mercury. With this method accurate pressure measurements are essential and in-pile instrumentation capable of withstanding the pressures involved is required for doubling time measurements and safety protection.

7.7.1. The Dynamic Technique for Control Rod Calibration by Air Poisoning

The method already described for the calibration of control rods by balancing at different air pressures is very time consuming and laborious. The dynamic technique is a modification of the method and enables the calibrations to be carried more easily.

The reactor is first balanced at a low power level of the order of 50 to 100 W. The air pressure is then increased at a slow continuous rate (of the order of 10 cm of mercury per hour). Simultaneously the control rods are withdrawn to maintain a balance. Hence a continuous calibration is obtained by plotting the rod position against air pressure. The increase in pressure and rod withdrawal are stopped at certain points for an air pressure coefficient measurement to be carried out following the procedure already described.

The advantages of this method are:

(a) A complete calibration can be obtained in the time taken to withdraw the control rods.

(b) A calibration is obtained at every rod position.

(c) In-pile instrumentation is not required as the installed equipment is operational at powers in excess of 50 W.

(d) The method is safer in that continual observation of the reactor power level is possible during the experiment.

7.8. THE MEASUREMENT OF SHUT-DOWN CAPACITY

The shut-down capacity is the negative reactivity of a reactor with all control rods and absorbers fully inserted.[14, 15] The safety of the reactor is dependent upon this parameter and therefore a measurement is required during commissioning and initial testing of the reactor system. Some of the methods already described for calibrating control rods may be used to obtain the shut-down capacity, the most frequently used being the rod drop technique. Three other methods are available and these are:

1. Method of count rate–reactivity matching.
2. Rod bank method.
3. Subcritical counting technique.

7.8.1. Count Rate–Reactivity Matching

This method involves subcritical and supercritical measurements. Absolute reactivity values are obtained from the supercritical measurements. These are used to convert neutron flux measurements in the subcritical region into equivalent values of the core reactivity.

With all control rods fully inserted the neutron flux is determined with a number of counters placed in or near the core. The control rods are then withdrawn in stages up to criticality and the reciprocal flux plotted against rod position.

The control rods are withdrawn in a number of stages above criticality and a doubling time measured at each stage. From these results the excess reactivity in the supercritical region is plotted against rod position.

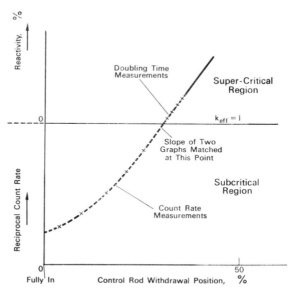

FIG. 7.3. Reciprocal count rate–reactivity matching

The reciprocal flux is converted to reactivity by adjusting the scale of the reciprocal flux against rod position graph until the slope of the subcritical plot is equal to the slope of the excess reactivity against rod position plot in the super-critical region. Hence the value of the reactivity with all rods fully inserted is the shut-down capacity of the reactor. The method is illustrated in Fig. 7.3.

The neutron flux may be converted to reactivity without the need for supercritical measurements provided that at some particular position of the rods in the subcritical region a reactivity determination is made by another method.

The multiplication M (see equation 6.1) is given by

$$M = A\phi = \frac{1}{1 - k_e} \tag{7.21}$$

where ϕ is the neutron flux,

k_e the effective multiplication factor, and

A a constant.

Hence, reactivity is given by

$$\varrho = \frac{k_e - 1}{k_e} = \frac{1}{1 - M} = \frac{1}{1 - A\phi} \tag{7.22}$$

By determining ϱ at a given position (ϱ_1) corresponding to a flux ϕ_1 by the rod drop method, then the constant A can be obtained from

$$A = \frac{\varrho_1 - 1}{\varrho_1 \phi_1} \tag{7.23}$$

Hence the subcritical flux measurements may be converted to reactivity values in order to obtain a value for the shut-down capacity.

7.8.2. **Rod Bank Method**

In this method the control rods are divided into a number of groups, each group being calibrated by a method such as the air poisoning technique. The total reactivity controlled is then obtained by adding the worths of the individual groups and applying a correction for the interference effects between the rods.

By adopting the air poisoning technique for this method reliable results can be achieved. The procedure to be adopted would be

1. Divide the control rods into a number of uniformly distributed groups. Ensure that the number of rods in each group enables criticality to be achieved with the group fully inserted at an air pressure of 25 cm of mercury.
2. Determine the total reactivity worth of each group by the air poisoning technique.
3. Divide one group of rods into two uniformly distributed subgroups each containing the same number of rods.
4. Determine the total reactivity worth of each subgroup by the air poisoning technique. Deduce the magnitude of the interference effects between the controls rods.
5. The total reactivity worth of all the control rods is obtained by summation of the worths of the groups and a correction applied for the interference effects.

7.8.3. **Method of Subcritical Counting**

The variation of neutron flux in the core with air pressure (or another distributed poison) is determined with all the control rods fully inserted and with two other uniformly distributed groups of rods. The total shut-down capacity may be deduced from these measurements.

The following procedure is adopted.

1. Select two groups of uniformly distributed control rods such that criticality is achieved at two different air pressures (e.g. 25 and 100 cm of mercury) and such that the radial flux distributions with the two groups and with all rods inserted is not dissimilar.

   Group 1 would be selected by reducing the vessel air pressure to 25 cm of mercury with all control rods inserted and then withdrawing rods one or two at a time. At the end of each withdrawal the reciprocal flux will be plotted against the number of rods left in the reactor. Hence by carrying out an approach to criticality in this way the exact number of rods in the core to achieve criticality at 25 cm of mercury pressure may be obtained. The procedure would be repeated with the vessel air pressure at 100 cm of mercury and the number of rods in Group 2 determined.

2. With all control rods inserted reduce the vessel air pressure to 25 cm of mercury.

3. Increase the vessel air pressure in stages (of approximately 15 to 20 cm of mercury) and determine the neutron flux in the core at each stage with all rods inserted and with Group 1 rods only inserted.

4. Continue the measurements and above 100 cm of mercury pressure determine the flux with all rods, with Group 1 rods and with Group 2 rods inserted.

5. Plot the reciprocal flux against air pressure for each of the three groups of rods.

6. Extrapolate the plot for all rods inserted to determine the total worth of the rods and hence the shut-down capacity.

This method of deducing the total worth of all the control rods is illustrated in Fig. 7.4 which shows the plots of reciprocal flux against air pressure for all rods, Group 1 rods and Group 2 rods inserted. The method is a subcritical method and the

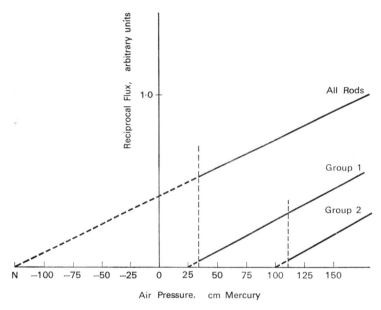

FIG. 7.4. Variation of a neutron flux with air pressure in sub-critical region

dotted vertical lines represent the limit of measurement at the two different pressures. Provided the groups of rods have been chosen so that the radial flux distribution across the core is unchanged for the three groups then the plots are straight lines which are parallel to one another. Hence although a large extrapolation of the plot with all rods in is required, this can be carried out reliably because the line can be drawn parallel to the other two plots. The plot for all rods in will cut the pressure axis at a negative air pressure (point N on Fig. 7.4). Multiplying this value by a previously measured value of the air pressure coefficient of reactivity will give the shut-down capacity of the reactor.

7.9. TEMPERATURE COEFFICIENT MEASUREMENTS

The temperature coefficient of a reactor is defined as the reactivity change per °C rise in temperature of the reactor.[2] Temperature coefficients are negative for water reactors and during the early stages of operation of graphite-moderated reactors. Consequently additional reactivity must be built into the core in order to overcome the negative reactivity effect of increasing temperature. An accurate measurement of the temperature coefficient at zero core irradiation is therefore required in order to improve design calculations for future projects. It is also necessary for the operators to have experimental information about the reactivity changes which are likely as a result of changes in temperature before commencing power operation.

In the case of the graphite-moderated reactors the value of the graphite temperature coefficient changes with irradiation due to the build-up of plutonium and the resultant change in the thermal neutron energy (see Section 9.8). The effect of this variation is to cause the combined (i.e. uranium and graphite) temperature coefficient to become increasingly more positive with core irradiation. It is therefore necessary from an operating point of view to continue to measure the temperature coefficient throughout the life of the reactor. These measurements at power will be dealt with in Chapter 9.

The measurement of the overall temperature coefficient at zero irradiation usually takes place on completion of the control rod calibrations. The experiments consist essentially of heating the core components to a number of temperatures and determining the resultant reactivity change from the control rod calibrations or by determining the excess reactivity of the core at each temperature.

The simplest method and the one usually adopted with water reactors is to balance the reactor at low power and determine the position of the control rods. Reactor temperatures are then raised in stages by heating the coolant by external means and a balance at low power achieved at each temperature. The change in reactivity with temperature is then determined from the control rod movement and the results of the control rod calibrations. With all temperature coefficient measurements it is important to achieve uniform reactor temperatures.

A more accurate temperature coefficient determination may be made by the uniform poisoning technique used for control rod calibrations. This technique is used during the commissioning of the gas-cooled reactors.

The reactor is first raised to a uniform temperature of approximately 60°C. The control rods are then withdrawn in stages and the vessel air pressure increased to absorb the reactivity released as the rods are withdrawn. Eventually a balance will be achieved with all the control rods completely withdrawn and the total excess reactivity absorbed by the air poisoning. With a number of control rods inserted, the air pressure is reduced slightly to a value such that with all control rods withdrawn the reactor flux will double with a period of between 80 and 100 sec. The control rods will then be withdrawn and the doubling time measured.

An air pressure coefficient of reactivity will then be determined by reducing the air pressure in two or three additional stages and again determining a doling time at each stage.

The whole procedure will be repeated with the reactor at uniform temperatures of 80, 100, 120 and 140°C. These temperatures are achieved by heat input from the gas circulators.

If the pressure at temperature T_1 for the first doubling time measurement is P_1 and the pressure coefficient is α_ϱ then the excess reactivity of the core is given by

$$k_e = P_1\alpha_\varrho + \Delta k_e \qquad (7.24)$$

where Δk_e is the free reactivity determined from a doubling time measurement. Hence at each temperature k_e is determined and the temperature coefficient deduced by measuring the slope of the resultant k_e against temperature graph.

7.9.1. Variation of Pressure Coefficient with Temperature

It is found that the air pressure coefficient is dependent upon temperature and consequently corrections are required to measurements of the air pressure coefficient.

An empirical relationship has been deduced and is given by

$$\alpha_2 = \alpha_1 \left(\frac{T_1}{T_2}\right)^n \qquad (7.25)$$

where α_2 and α_1 are the pressure coefficients at temperatures T_2 and T_1.

The value of n varies between 1·00 (assuming the gas behaves in accordance with Charles' law) and 1·25 (assuming Boyle's law is applicable).

7.9.2. Hot Control Rod Coefficient

On account of the dependence of neutron cross-sections on neutron temperature the worth of a system of control rods varies with temperature. The hot control rod coefficient is defined as the change in reactivity worth per °C rise in control rod temperature. Control rods are calibrated at ambient temperatures and therefore a measurement of the hot control rod coefficient is desirable in order to correct the calibrations for operating temperatures.

The hot control rod coefficient is determined by measuring the worth of different groups of control rods at different temperatures by the air poisoning technique (or for water reactors the distributed poison technique).

Let us assume that the balancing pressure with a group of rods of reactivity worth k_r is P and that the balancing pressure difference with this group of rods fully inserted and fully withdrawn is Δ_p.

Then, at temperature T_1

$$P_1\alpha_1 = k_{e_1} - k_{r_1}$$

and

$$k_{r_1} = \Delta P_1\alpha_1$$

where α represents the air pressure coefficient,

k_e the total excess reactivity of the core, and subscript 1 refers to temperature T_1.

At temperature T_2

$$P_2\alpha_2 = k_{e_1} - \alpha_T(T_2 - T_1) - k_{r_1}[1 + f(T_2 - T_1)]$$

where α_T represents the temperature coefficient,

f the hot control rod coefficient, and subscript 2 refers to temperature T_2.

Hence

$$\frac{P_1\alpha_1 - P_2\alpha_2}{(T_2 - T_1)} = \alpha_T + k_{r_1}f$$

and using $\alpha_2 = \alpha_1\left(\dfrac{T_1}{T_2}\right)^n$ we get

$$\frac{P_1 - P_2\left(\dfrac{T_1}{T_2}\right)^n}{(T_2 - T_1)} = \frac{\alpha_T}{\alpha_1} + f\Delta P_1 \qquad (7.26)$$

By determining the balancing pressure P and the pressure difference ΔP at two different temperatures (60°C and 140°C) for a number of groups of rods it is possible to plot $P_1 - P_2 \left(\dfrac{T_1}{T_2}\right)^n (T_2 - T_1)$ against ΔP_1. The slope of this graph will give f, the hot control rod coefficient.

REFERENCES

1. KEEPIN, G. R. *Physics of Nuclear Kinetics*, p. 166 and Section 8.5. Addison-Wesley (1965).
2. GLASSTONE, S. and SESONSKE, A. *Nuclear Reactor Engineering*. Van Nostrand (1964).
3. STURM, W. J. (ed.). Reactor laboratory experiments. *Argonne Nat. Lab. Report* ANL-6410.
4. SCHULTZ, M. A. *Control of Nuclear Reactors and Power Plants.* McGraw-Hill (1955).
5. PEREZ-BELLES, R. *et al. Nucl. Sci. and Eng.* **12**, 505 (1962).
6. RUANE, T. F. *Nucleonics* **20**, 10 (1962).
7. HOGAN, W. S. *Nucl. Sci. and Eng.* **8**, 518 (1960).
8. JANKOWSKI *et al. Nucl. Sci. and Eng.* **2**, 288 (1957).
9. FEINER, F. *et al.* Pile oscillator techniques. *U.S.A.E.C. Report* KAPL-1703 (1956).
10. BENGSTON, J. *Proc. of Second Geneva Conf.* **12**, 84 (1958).
11. SIMMONS, B. E. *Nucl. Sci. and Eng.* **5**, 254 (1959).
12. SAUSSURE, G. D. *et al. Nucl. Sci. and Eng.* **9**, 291 (1961).
13. SAUSSURE, G. D. and SILVER, E. G. *U.S.A.E.C. Report* ORNL-2641 (1959).
14. *Transactions of American Nuclear Society* 1960-1966. Sections
    (a) Reactivity and subcriticality measurements;
    (b) Reactivity measurements;
    (c) Pulsed neutron measurements.
15. MURRAY. R. L. *Nuclear Reactor Physics.* Prentice-Hall (1957).

CHAPTER 8

Fuel Element Leak Detection and Fuel Movement

8.1. INTRODUCTION

Fissile material used in nuclear reactors is canned or sheathed in materials which have structural stability and are compatible with the fissile fuel. Bare fuel can not be used because the fission products released during operation will enter the coolant and produce widespread contamination with a resultant radiation hazard. In addition rapid corrosion of uranium occurs in the presence of the coolant (gas or water) which can lead to considerable breakdown and deformation of the fuel elements. A variety of materials are used to can uranium and other fissile fuel in reactors. Pure aluminium, alloys of aluminium, magnesium and zirconium and stainless steel are the most frequently used on account of the fact that they are compatible with uranium and are therefore less likely to cause rapid corrosion. Stainless steel has a disadvantage in that it has a high neutron absorption cross-section and therefore effects the neutron economy of the reactor. It is used in high-temperature reactors when a high degree of structural stability is required.

In the event of a fault occurring in the canning material due to a structural defect or corrosion then a leak of fission products may occur which would cause contamination. Consequently it is usual to provide equipment to detect fuel element leaks at an early stage of development. The faulty fuel element is then discharged from the core before any serious damage occurs.

Many methods have been considered for the detection of faulty fuel elements.[3] In small reactors it is usual to attempt to detect the presence of fission products in the coolant stream and then to examine each element in order to locate the fault. In large power reactors the problem is a different one. An attempt in this case is made to detect a leak in an individual fuel channel and therefore complications arise due to the large numbers of channels involved (more than 3000 in a graphite-moderated, gas-cooled reactor).

Some non-nuclear methods of detection have been considered. For instance, the physical deformation of a can due to a large leak could be detected by the change in coolant flow as a result of the deformation. Obviously considerable damage would already have occurred before detection in this case. Other methods involve either pressurising each can and detecting the fall in pressure when a leak occurs or sealing a detectable fluid or gas within the can which is released as a result of a fault. In both cases considerable instrumentation is required and the leak may be missed altogether due to faulty equipment. Without doubt the best methods depend upon the detection of fission products within the coolant. These may be detected by four methods.

(a) Gamma activity of the fission products.
(b) Neutron emission from certain fission products.
(c) Beta emission from fission products.
(d) Gamma and beta activity of gaseous fission products and their daughters.

Before discussing each of these methods in detail, some consideration will be given to the most favourable systems used on different types of reactors. As already mentioned, the problem is usually one of distinguishing the activity of particular fission products from a general background of activity. The selection of fission products which emit delayed neutrons has been carried out for the PWR reactor.[1] At Savannah River[2] a system has been used which separates fission-product xenon and krypton from the reactor cover gas by gas chromatography. With most gas-cooled reactors the wire precipitator method is used to detect xenon and krypton in the coolant stream.[1, 5] In boiling water reactors monitoring of the main turbine gas stream is carried out for detection of gaseous fission products.

The location of failed elements among the many elements of the core is a much more difficult task as a large amount of in-core equipment may be necessary to monitor individual elements of individual channels. In tube-type or channel type reactors it is fairly straightforward to identify the individual tube or channel containing the failed element if all the tubes are sampled individually and if the fission product detected has a half-life which is short compared to circulation time of the reactor coolant.

In most water reactors the elements are not usually well segregated and the installation of an on-power detection system is very complex. On-power detection in water reactors has only been partly successful so far. Normally in this case individual elements are detected by withdrawing suspects into a test container to check for build-up of fission products.

In water-moderated reactors using uranium oxide fuels an effect has been observed which provides an efficient system for failed element detection. Defective elements emit puffs of fission product gases when the power in the elements is increased rapidly after a period of shut-down. Water enters the element at shut-down and leaches some fission products from the oxide fuel. These products are expelled with the water when power is increased and the puff may be detected. This system has been adopted in the Shippingport reactor.[1]

8.2. DETECTION BY FISSION PRODUCT GAMMA ACTIVITY

The presence of fission products in the coolant may be detected merely by measuring the gamma activity of the coolant. This is not a very satisfactory method if a high degree of sensitivity is required on account of the large gamma background due to activated impurities in the coolant. In this case a leak would pass undetected until the emission of fission products was large enough to indicate a gamma activity in excess of the normal background. The sensitivity can be improved by a method in which the activity of the fission products alone is determined. This may be done in water reactors by the use of an ion exchange column placed in the coolant stream. Ion exchange involves the removal of cations or anions (or both) from the water by adsorption into an organic resin. The resin releases in exchange the H^+ or OH^- ions. Hence the gamma activity of the exchange column will build up due to the capture in the column of the fission products. Reasonably efficient detection can be achieved by measuring an integrated activity over a certain time period. This method is used extensively in low-power water reactors.[4]

The system may also be used to monitor a reactor with multiple coolant channels. An intricate system of selector valves may be used to pipe a sample of coolant from each coolant channel (similar system to that described in Section 8.6) into a resin detector column. The integrated activity of the column is then determined.

8.3. DETECTION BY DELAYED NEUTRON EMISSION

Neutrons are emitted by the fission products bromine-87 and iodine-137 with half-lives of 54·3 and 21·7 seconds respectively. By using a BF_3 or scintillation counter the neutrons may be detected. The transit time of the coolant between the leaky fuel element and the detector is important in this method. The time must be comparable with the neutron emission half-life so that a reasonable count rate is obtained. The method is suitable for water reactors.[5]

The main disadvantage arises as a result of the presence of the isotope oxygen-17 in carbon dioxide or water. Neutron bombardment of oxygen-17 produces the reaction O-17 (n, p) N-17. The nitrogen-17 isotope emits a 1 MeV neutron with a half-life of 4·4 sec. Consequently the overall sensitivity of the method is reduced due to this effect.

The efficiency of the delayed neutron method may be increased considerably by the use of isotopic exchange. Isotopic exchange is a process which occurs with silver bromide or silver iodide in which the isotopes of bromine or iodine are exchanged in accordance with the reaction

$$AgX + X' = AgX' + X$$

where X and X' represent two different isotopes. Therefore, if a sample of water coolant from each reactor channel is passed through a column containing silver iodide then there will be a build-up of the isotopes of bromine-87 and iodine-137 in the column when a fuel element leak occurs. The neutron emission from the isotopic exchange column is detected.

8.4. DETECTION BY BETA EMISSION

The presence of fission products in the coolant of water reactors may be detected using a Cerenkov detector.[6] High energy electrons are emitted from fission products. It is known that electrons traversing a transparant medium emit light provided that the velocity in the medium is greater than the velocity of light in that medium. The visible radiation emitted in this manner is known as Cerenkov radiation. If the water coolant is allowed to pass a photomultiplier then pulses will be produced which are proportional to the amount of Cerenkov radiation emitted and therefore indicate the presence of fission products. The system easily discriminates from operational backgrounds as only high energy electrons (of order 5 MeV) are detected and these are only emitted by fission products and not activated impurities. The advantage of the system is therefore that it can operate satisfactorily in very high backgrounds.

8.5. DETECTION BY GASEOUS FISSION PRODUCT ACTIVITY

A number of radioactive gaseous fission products are produced during the fission process and these may be used in a variety of methods for the detection of leaky fuel elements. A large number of isotopes of krypton and xenon occur as a result of radioactive decay. These isotopes decay with beta emission to form isotopes of rubidium and caesium which are also beta active. The advantage of this particular method is that the daughter products rubidium and caesium are solids. Therefore by careful filtering it is possible to extract the gases krypton and xenon from other impurities in the coolant stream and allow them to decay to solid daughter products. The beta activity of the daughters is then measured in order to detect the presence of fission products.[7, 9]

The gaseous fission products in a stream of gas are usually allowed to enter a precipitation chamber where the solid daughter products are produced. These daughters may be collected on a central wire in the precipitation chamber by maintaining a high voltage between the chamber and the wire. The activity collected on the wire is then measured. The method is described in detail in Section 8.6.

The gaseous fission products may easily be obtained from the coolant stream in a gas-cooled reactor by passing a small amount of the gas from each fuel channel to the precipitation chamber. The problem is more difficult in water reactors.[8] In this case the gas sample is obtained either from a gas blanket in the pressure vessel or from a deaerator in the water stream. A deaerator is provided to remove entrained air and radiolytic gases from the coolant. Gas samples obtained in this way are then passed to a precipitation chamber for fission product detection.

The activity of the solid daughter products of the gases krypton and xenon may also be detected by the method of gas chromatography.[10, 11] A gas chromatograph employs the principle that gases of different atomic weights present in a gas sample move at different speeds through a molecular sieve column and therefore different gases may be separated. The gas chromatograph accepts a gas sample and injects it into the sieve column which sequentially rejects other gases but passes krypton and xenon into a sampling chamber. The activity in the sampling chamber is detected by a scintillation counter. This method yields a greater sensitivity than the precipitation method for water reactors in view of the greater gaseous collection efficiency. In the case of gas-cooled reactors the precipitation method is by far the best and a very high degree of sensitivity may be achieved.

The most frequently used method for fuel element leak detection is that of the precipitation method and therefore this technique will be described in some detail with particular reference to the gas-cooled reactor system. The principles employed are applicable to other types of reactor and only differ in the method of obtaining a gas sample.

8.6 THE WIRE PRECIPITATOR METHOD

8.6.1. The Sampling System

A typical gas-cooled power reactor fuelled with natural uranium will have over 3000 fuel channels. Obviously it would be quite uneconomical to monitor each channel for the presence of fission products continuously and therefore a compromise is adopted. A small pipe is fitted at the top of each channel in such a way that a gas sample is picked up from each channel. A flow rate along the sampling pipe of approximately 1% of the channel flow rate is an adequate amount. The sampling pipes are brought out of the reactor pressure vessel and the shield and are connected together into groups. The combined gas sample from these groups is then passed to a selector valve controlling a large number of groups. This valve selects each group of channels in sequence and allows the gas sample to pass to a precipitation chamber.

Before the gas sample enters the precipitation chamber it is cooled and filtered in order to remove particles of dust and other solid fission products from the gas stream. This is necessary in order to ensure that the system detects the daughter products of krypton and xenon, and to prevent contamination of the precipitation chamber and detector unit which would lead to a high background activity and a corresponding reduction in overall sensitivity. The filter will remove from the system any rubidium and caesium isotopes which have been produced before the krypton and xenon have passed through the filter. This will again reduce the sensitivity of the system. The gas sample from one particular group of channels is allowed to flow through the precipitation chamber for a sufficient period to allow the solid daughter products of krypton and xenon to be collected. The gas flowing out of the precipitation chamber is then returned to the main gas circuit or pressure vessel via a compressor to maintain the pressure of the gas circuit.

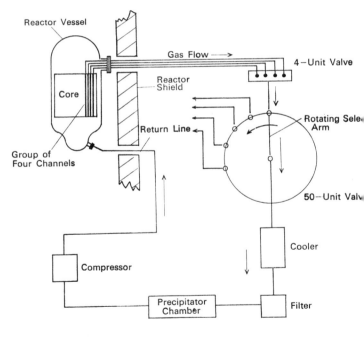

Fig. 8.1. Gas sampling arrangement

The system is shown schematically in Fig. 8.1.

The purpose of the selector valve is to reduce the number of precipitation chambers required to monitor the complete reactor core. If a large number of precipitators are used then only a few valve selector positions would be required and a scan of the reactor would be very rapid. Such a system would be expensive. In practice it is found that fuel element leaks occur gradually and are slow to develop in the initial stages. A period of 30 min is considered adequate for a complete scan of all the groups of channels. Therefore if we assume that there are 3200 channels in the reactor core and that these are connected into groups of four, there will be 800 groups. Such a system could be monitored using 16 precipitation chambers and 16 selector valves, each valve having 50 positions. The selector valve is usually a rotary valve which allows the gas from a particular group to be sampled and then moves automatically to allow the next group to be sampled. Hence, with this system the complete reactor will be scanned after the rotary selector valves have moved to each of the 50 positions. Assuming a sampling time of 25 sec, then a complete scan will occupy a period of just over twenty minutes.

One or two additional spare precipitation chambers are usually provided so that continuous monitoring of a suspect group may be carried out. In the event of fission products being detected in the gas sample from a group of channels the sample will pass to a spare precitator and each individual channel in the group scanned to detect the channel containing the fault.

It is also common practice to provide one precipitator to monitor gas samples obtained from the main coolant outlet and inlet ducts. These will provide an indication if a very large leak occurs in a fuel element sufficient to increase the fission product activity significantly throughout the whole coolant.

8.6.2. The Precipitation Chamber and Detection System

The precipitation chamber consists of a metal cylinder with a thin metal wire passing through its centre insulated from the cylinder. The cylinder potential is approximately 4 kV positive with respect to the wire. The gas from each sample passes through the cylinder for a period of approximately 25 sec and during this time the solid daughter products rubidium and caesium are collected. These isotopes are positively charged and therefore migrate to the central wire and become attached to it. At the end of the sampling period the wire is moved rapidly out of the chamber and into a hole in a phosphor and photomultiplier system. A beta-sensitive phosphor is used and the photomultiplier output is amplified to operate a ratemeter. The phosphor is usually shielded with lead to reduce the background as much as possible. The presence of beta active materials on the wire is detected by determining the integrated count over a period equal to the gas sampling period (i.e. approximately 25 sec).

The precipitator wire passes over a system of pulleys and back through the precipitator chamber. Sufficient time must elapse before a particular part of the wire re-enters the chamber to allow for the complete decay of the deposited active products. The half-lives of rubidium and caesium isotopes are of the order of minutes and it is found that their activity has reduced by a factor of 10 in approximately 20 min. A reduction factor of 100 is achieved by making the wire about 70 ft in length; this wire being coiled on a specially designed storage drum.

A background signal is produced in the detection system due to the presence of argon-41 which emits a beta ray in the carbon dioxide gas stream. In order to reduce this effect it is usual to purge the phosphor continually with clean inactive carbon dioxide which then passes into the gas stream.

The complete detection system is represented schematically in Fig. 8.2.

Many different, more sophisticated, designs of precipitators are now used for detection of fission products. Most are more elaborate versions of the wire method described. A different

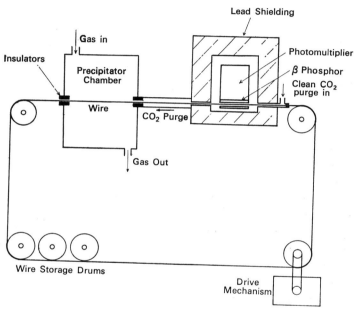

FIG. 8.2. Precipitator chamber and detection system

system has been used at the Chinon Power Station in France.[12] The detectors in this case consist of a rotating drum, a segment of which forms one wall of a precipitator chamber. The active solid products are collected on this wall and the activity measured using scintillation counters and scalers. The deposited activity is then stripped from the wall of the drum by reversing the polarity of the potential on the wall. Two scalers are associated with the detector. One measures the total gaseous activity and the other the gaseous activity plus the precipitated activity. A computer is used to take the difference and display the activity of the precipitated products.

During operation a complete scan of 3200 channels can be obtained in a period of approximately 20 min. This process occurs continuously and consequently a large amount of data must be recorded. Recording of data may be carried out by analogue or digital methods. The ratemeters associated with each precipitator produce a signal proportional to the count integrated over the sampling period. This signal may be amplified to operate a multi-point recorder for direct display of the measured activity. Recorders have been developed to record the activity from as many as 50 to 60 samples. An alternative system uses the output from the amplifiers to operate digital circuits which in turn operate electric typewriters to print out the activity measurements.

Warnings are usually incorporated in the display and recording equipment to indicate when any particular signal exceeds a pre-set limit. When this occurs the operator will investigate the situation and display the measured activity from the particular sample continuously on a spare precipitator.

At the end of a sampling period the following separate operations are carried out automatically:

(a) The selector arm of the rotary valve is advanced by one unit.

(b) The precipitator wire is snatched out of the precipitation chamber and moved into the detector head.

(c) The total integrated count as measured by the ratemeter is recorded.

(d) The recorder moves to the next position.

(e) The ratemeter is re-set.

8.6.3. Performance of the Wire Precipitator System

The performance of the precipitation system is governed by the following parameters:

(a) Reactor power.
(b) Coolant mass flow.
(c) The number of channels in each monitored group.
(d) The coolant mass flow through the precipitation chamber.
(e) The precipitation chamber gas pressure, voltage and volume.
(f) The transit time for the gas to pass through the precipitation chamber.
(g) The gas sampling time.

The maximum performance is obtained by optimising these parameters.[1] The relative effects on the performance will be considered.

Assume that the gas entering the precipitator chamber contains N_x atoms per gram of the gaseous fission product xenon at any time t. This will decay in accordance with the exponential radioactive decay law such that

$$\frac{dN_x}{dt} = -\lambda_x N_x \qquad (8.1)$$

where λ_x is the disintegration constant for a particular xenon isotope.

Xenon decays to caesium and therefore the rate of formation of caesium atoms is given by

$$\frac{dN_c}{dt} = \lambda_x N_x - \lambda_c N_c \qquad (8.2)$$

where λ_c is the disintegration constant for the appropriate caesium isotope and N_c is the number of caesium atoms at any time t.

These two equations express the fact that xenon decays according to the radioactive decay law and that caesium atoms are formed at the rate of $\lambda_x N_x$ because of the decay of xenon and disappear at the rate of $\lambda_c N_c$.

The number of xenon atoms at any time t is given by

$$N_x = N_x(0) \exp(-\lambda_x t) \tag{8.3}$$

where $N_x(0)$ is the number of xenon atoms at time $t = 0$. Substituting in equation (8.2) we may write

$$\frac{dN_c}{dt} + \lambda_c N_c = \lambda_x N_x(0) \exp(-\lambda_x t) \tag{8.4}$$

On integration this gives

$$N_c \exp(\lambda_c t) = \frac{\lambda_x}{\lambda_c - \lambda_x} N_x(0) \exp(\lambda_x - \lambda_c)t + C \tag{8.5}$$

where C is a constant of integration.

The value of C may be obtained by assuming that at time $t = 0$, $N_c = N_c(0)$ which is constant.

Then substituting in equation (8.5) and rearranging we have

$$N_c = \frac{\lambda_x}{\lambda_c - \lambda_x} N_x(0)[\exp(-\lambda_x t) - \exp(-\lambda_c t)]$$
$$+ N_c(0) \exp(-\lambda_c t) \tag{8.6}$$

Hence if the number of xenon atoms entering the precipitator chamber at the commencement of a sampling period is $N_x(0)$ per gram and it is assumed that initially there are no caesium atoms present (i.e. $N_c(0)$ is zero); the number of caesium atoms per gram formed during the passage of gas through the chamber is

$$N_c = \frac{\lambda_x}{\lambda_c - \lambda_x} N_x(0)[\exp(-\lambda_x t_c) - \exp(-\lambda_c t_c)] \tag{8.7}$$

where t_c is the transit time for the gas to pass through the chamber.

If the CO_2 flow rate through the precipitator is f grams per second, then the rate of formation of caesium atoms is fN_c per second.

Gas is allowed to flow through the precipitator for a period long enough to allow build-up of activity on the wire. This period is known as the sampling period t_s. During this time caesium will itself decay and the net rate of formation of caesium atoms may be expressed as

$$\frac{dN}{dt} = fN_c - \lambda_c N \qquad (8.8)$$

On integrating and using the fact that $N = 0$ at $t = 0$ we obtain

$$N = \frac{fN_c}{\lambda_c}(1 - \exp(-\lambda_c t_s)) \qquad (8.9)$$

This represents the number of radioactive caesium atoms collected during the sampling period t_s. For maximum performance of the system this value must be as large as possible.

The value of N may be increased by increasing the gas flow f. However, the transit time t_c is inversely proportional to the gas flow and N_c will be reduced if t_c is reduced (see equation 8.7). As can be seen from equation (8.9) the sampling period t_s needs to be such as to allow a reasonable deposition of radioactive atoms on the wire.

Many different radioactive isotopes of krypton and xenon are produced during fission. However, only about ten isotopes have appreciable yields and suitable half-lives for consideration in this context. The main contributors have half-lives in the range of 5 to 20 sec (i.e. krypton-91—10 sec and xenon-140—16 sec). The half-lives for the decay of the daughter isotopes vary in the range 60 to 150 sec (i.e. rubidium-91—100 sec and caesium-140—66 sec).

The coolant flow in the reactor is fixed by the design of the core. The flow rate through the sampling equipment may be altered by using a suitable system of valves or a compressor. Long *el al.*[3] have pointed out that ideally the transit time for the gas sample to pass through the chamber should equal the time for an ion to migrate from the outer edge of the chamber to the wire. This would ensure the most efficient performance. The transit time through the chamber and the migration time are dependent upon the chamber voltage, volume and gas pressure. The transit time in the chamber is given by

$$t_c = \frac{d\pi L R^2}{f} \tag{8.10}$$

and from electrostatic theory the migration time t_m is given by

$$t_m = \frac{930 \, R^2 \log_n \dfrac{R}{r}}{2 V \mu} \tag{8.11}$$

where d is the gas density,
L is the axial length of the chamber,
R is the radius of the chamber,
r is the wire radius,
V is the chamber voltage, and
μ is the ionic mobility.

The optimum voltage is obtained by equating equations (8.10) and (8.11) and is

$$V = \frac{930 f \log_n \dfrac{R}{r}}{2\mu \, dL\pi} \text{ volts} \tag{8.12}$$

In practice the chamber voltage is set at approximately 3 to 4 kV and the transit time is of the order of 20 to 30 sec.

The sampling period is governed by the rate of decay of the rubidium and caesium isotopes. The half-lives in this case are of the order of 1 min. Sampling periods are usually slightly longer than the transit time to allow gas to pass completely through the system during the sampling period.

The number of sampling pipes connected together to form the groups for monitoring is dependent upon the degree of dilution of a gas sample from a faulty channel. Increasing the number of pipes connected together will decrease the time required to monitor the whole reactor. This would reduce the possibility of missing a rapidly developing fault during a particular scan. However, if too many pipes are connected then the number of collected active isotopes for a given fault would be reduced with a consequent reduction in performance. It is common practice to connect the sampling pipes into groups of four or six.

8.6.4. Sensitivity and the Selection of Fault Criteria

The sensitivity of a fuel element leak detection equipment is determined in terms of the minimum area of exposed uranium which it is possible to detect. The presence of fission products in the coolant stream is measured in terms of an integrated count of the activity deposited on the wire. The problem is to correlate the count obtained with some quantity dependent upon the magnitude of the fuel element can fault. It is usual to calibrate the detection equipment against a known area of exposed uranium (Section 8.6.5). Thus it is possible to specify faults which produce the same effect as an equivalent area of exposed uranium. The physical size of a fuel can fault cannot be correlated with respect to an equivalent area of uranium on account of the diverse way in which faults develop. However, defining sensitivity in this way provides a means of specifying the performance and effectiveness of the equipment in a quantitative manner.

The measured count integrated over the sampling period is determined by:

(a) The surface contamination of the fuel elements. Despite elaborate precautions a certain amount of bare uranium is always present on the surface of the fuel can. This occurs during manufacture and is unavoidable. Approximately 10 to 20 μg/ft can be expected. During reactor operation a number of fission products will enter the gas stream from this contaminant and will be detected with the precipitator equipment.

(b) Background radiation.
A background will be present in the counting system and will make a contribution to the integrated count.

(c) Coolant activity.
Despite filtering a certain number of active impurities will enter the precipitator chamber and become attached to the collecting wire. These will add to the background count.

(d) Activity of the deposited rubidium and caesium isotopes. For the greatest sensitivity the maximum signal to background ratio (R) is required. This may be expressed as

$$R = \frac{C_f}{C_b + 4C_s + C_c} \qquad (8.13)$$

where C represents the count rate and the subscripts refer to the background (b), the surface contamination (s), the coolant activity (c), and the fission products (f).

The contribution due to surface contamination is $4C_s$ assuming that C_s is the contamination per channel and that four channels are combined into one monitored group.

To obtain the maximum value of R the values of C_b, C_s and C_c must be reduced to a minimum. The main contribution is due to C_s and it is found that C_b+C_c is less than 10% of the magnitude of C_s. In practice it is found that the value of $4C_s$ is of the order of 500 counts in a 25-sec sampling period.

The criteria to be adopted in order to decide whether a faulty fuel element has been detected depends on local circumstances. However, warnings are usually set to draw the operators attention to a group of channels when the integrated count exceeds a value of twice the normal operational count. In this case the particular channel in the group would be selected and observed continuously. It is desirable to remove the faulty element from the reactor in the event of the integrated signal exceeding a value of ten times the normal operational background.

The sensitivity of the system to faults of a particular magnitude is reduced during operation due to the general increase in contamination of the coolant channels and the leak detection equipment. Hence the value of C_s is continually increasing. If a leaky can does develop in a channel a certain amount of additional contamination in that channel is unavoidable. Similarly a rapid burst may cause considerable contamination before shut-down and removal are effected. Thus the faulty element selection criteria must be continually reviewed throughout the life of the reactor. It may be that for groups of contaminated channels with a high value of $4C_s$ a fault is assumed to have occurred with only a 30% increase in the observed count.

8.6.5. The Calibration of the Detection Equipment

The sensitivity of the equipment to fuel element leaks in any part of the reactor is determined quantitatively by introducing foils of bare uranium into the reactor. These tests are carried out during the commissioning and are usually the last test prior to commencing operation at power. The minimum area of exposed uranium which can be detected is found by measuring the counts obtained as a result of the irradiation of known areas of uranium. In addition the effects of variations in precipitator voltage, reactor power and coolant flow are obtained enabling the optimum operating conditions to be established.

To avoid excessive irradiation of the core which would result in fuel element handling difficulties at the end of the tests, it is essential to perform the experiment at a low power level. Consequently as the fission product signal to be detected is proportional to power then comparatively large areas of exposed uranium are required to obtain satisfactory results.

Foils of bare natural uranium or enriched uranium are used to simulate leaking fuel elements. The use of enriched uranium will enable smaller foils to be used. In this case the foils must be calibrated in terms of their equivalent area of natural uranium. The method of securing the foils in the coolant channels is such that the gas stream flows past both sides of the bare foil hence increasing the total exposed area. The actual method of attaching the foils is dependent on the physical arrangement of the core. It is often convenient to attach the foils to the fuel elements themselves. In this case due consideration must be given to the fuel-handling problems associated with foil removal at the end of the experiment. Excessive radiation doses to operating personnel must be avoided.

To obtain a comprehensive calibration of the detection system uranium foils should be positioned such that results are obtained for differing physical parameters. As the lengths of the sampling pipes between the channels and the precipitators vary then the transit time for the gas to pass to the precipitator will vary. Hence a variation in integrated count will result due to these differences. The change in sensitivity due to this effect may be measured by the suitable positioning of the foils. A further cause of a decrease in sensitivity will be due to incomplete mixing of the gas stream with the fission products. The worst possible case will occur when a leak occurs in a fuel element nearest to the nostril of the sampling pipe but on the opposite side of the channel. Foils are positioned to investigate this effect. Finally the effects of leaks occurring at different positions in the channels are also investigated.

The integrated counts obtained from the channels containing the bare foils is dependent on the neutron flux at the foil. The greater the flux the greater will be the emission of fission products. To obtain a true calibration therefore the flux at the uranium foils must be determined during the tests. This is done by attaching manganese foils or tungsten wires to the foil and determining the flux by activation techniques.

The experimental procedure consists of loading the foils, adjusting the gas flow (air is usually used during the tests) to the normal operational level and then raising the reactor power. The reactor power for the tests is determined by observing the integrated count as measured with the detection equipment for a channel in a position of low neutron flux as the power is raised. The power will be balanced at a level such that a reasonable count is obtained from this channel. A value of 50% to 100% above the background is an adequate amount. The power will be held steady at this level and the equipment allowed to carry out three or four complete scans of the whole reactor. In addition the signals obtained from individual channels in the groups containing a simulated leak will be determined.

Finally the effects of reducing the reactor power and varying the gas flow can be investigated.

In practice it is found that to obtain adequate integrated counts in the detection equipment an exposed area equivalent to approximately 1000 cm² of natural uranium are required and a reactor power level of the order of 10 to 50 kW (flux of order 10^9 n/cm² sec).

The results may be extrapolated to deduce the minimum area of exposed uranium which may be detected by the system. Corrections must be applied for the differences in neutron flux, temperature and coolant flow between the test conditions and normal operation.

As mentioned previously, samples are often obtained from the coolant stream in the inlet and outlet ducts to the reactor. Obviously if it is necessary to calibrate this part of the system much greater fission product release must be achieved on account of the large amount of dilution. In this case adequate signals have been obtained with 20,000 cm² of exposed natural uranium and a reactor power level of 100 kW. If these tests are performed arrangements must be made to remove the uranium foils from outside the reactor by remote handling techniques due to radiation hazards.

8.7. FUEL MOVEMENT

An essential part of the operation of reactors is the movement and handling of fuel elements. Provision must be made to charge the reactor core with new fuel, to discharge irradiated fuel and in some cases to carry out recycling of the fuel.[13, 15] In the case of research reactors it is necessary to be able to discharge the fuel for maintenance purposes or for the removal of spent fuel or damaged fuel elements. A fuel store is usually provided and arrangements must be made to transfer the fuel from the core to the store and vice-versa. In power reactors spent fuel or damaged fuel elements are removed and stored prior to transporting to a fuel processing plant.

The problems associated with the charge and discharge of fuel are concerned with the protection of operating personnel from exposure to high radiation levels, the prevention of the release of fission products and the prevention of serious contamination.[14]

Fuel-handling equipment is complex and elaborate as large numbers of different functions have to be carried out. It is an advantage in power reactors to be able to carry out charge and discharge with the reactor operating at full power. Hence a system of valves is required to allow access to a fully pressurised system. Research reactors and power reactors with small cores are shut down for charge and discharge of the fuel.

Complex machines are provided to remove fuel from the reactor core. To allow access to the core channels openings into the reactor vessel are required. These are usually provided at the top of the core, and stepped plugged tubes are provided allowing access through the top shield. Arrangements have to be made to withdraw the plugs, to open the tubes, insert guide mechanisms for positioning the fuel, charge and discharge the fuel by remote means and replug the hole without an undue radiation hazard and without releasing any coolant from the core. These operations are sometimes carried out with different machines, each performing one function, and sometimes with one multi-purpose machine. The active fuel elements must then be transferred to a suitable storage facility for subsequent removal.

To illustrate the problems involved, a brief description will be given of the method of charging and discharging a typical gas cooled reactor enclosed in a steel pressure vessel. The method is illustrated in Fig. 8.3.

The charge/discharge machine is first located above the appropriate plugged opening into the vessel. The plug is stepped to prevent radiation leakage. The plug (A in Fig. 8.3) is removed and stored in the machine. A fuel element guide tube (B) is inserted into the core and located in a fixing point

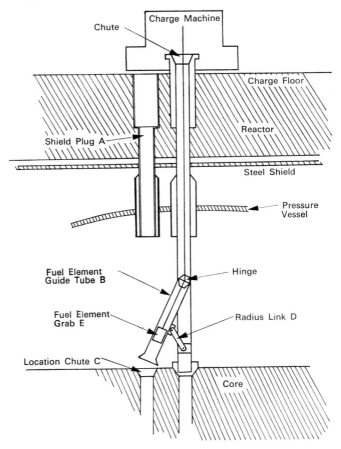

FIG. 8.3. Typical charge/discharge arrangement

on top of the core itself (C). The guide tube may be manipu-
lated remotely with controls on the charge machine so that a
chute attached to the guide tube may be lifted by a radius
link (D) to scan a number of different channels. By rotating
the whole guide tube then a large number of channels may
be located with the charge chute. In this way a total of 64
channels may in fact be located through one opening into
the pressure vessel. Figure 8.4 indicates how the channels

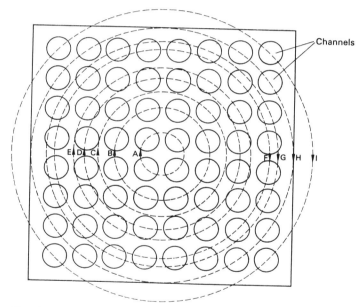

Fig. 8.4. Charge chute radial location system for a group of 64
channels

may be located with 9 radial settings of the chute. With the
charge chute located above a particular channel a remotely
operating grab (E) is lowered on a cable into the channel and

picks up a fuel element. This is withdrawn into the charge machine and placed in position in a loading basket. The process is repeated and fuel removed as required or until the loading capacity of the machine is attained. The machine is removed from the opening and the basket containing the elements placed into a container for removal to the fuel store. This is often done via an underground or heavily shielded channel.

Fuel stores are usually in the form of open water ponds, the elements being stored in a sequential arrangement. A water depth of 25 to 30 ft is required to provide adequate shielding. Fuel is normally stored on the reactor site in the ponds for a period of 3 to 9 months to allow the activity to decay partially. After this period the fuel will be loaded into shielded containers for transportation to a processing plant.

A normal fuel-handling cycle is represented schematically in Fig. 8.5. Fuel arrives on site and is stored in a "new fuel" store. It is then taken to the fuel element preparation room and each element is inspected visually for defects. The elements are then moved to a charging room near to the reactor charge face and are loaded into the machine baskets. The baskets are hoisted into the machine and charged into the reactor. After irradiation the active elements are discharged, stored and eventually removed from the site.

Additional remote handling facilities are also provided to withdraw, store and replace control rods, to remove damaged fuel elements and to visually inspect the core with remotely operable television cameras.

In view of the complexities of the charge and discharge procedures the time taken for the removal of a channel of fuel elements is lengthy. A period of 1 hr is common for the removal of a single channel.

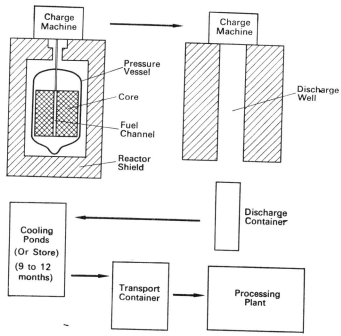

Fɪɢ. 8.5. Typical fuel-handling cycle

REFERENCES

1. Aʙʀᴀᴍ, S., Fʀᴀɴᴋ, P. W., Pᴀʀʀ, O. D. and Vᴀᴜɢʜɴ, F. R. Evaluation of the PWR core. Failed element detection and location system. Report WAPD – TM – 330 Bettis Atomic Power Laboratory (1962).
2. Kʀɪᴛᴢ, W. R. An automatic gas chromatograph for monitoring of reactor fuel failures. Savannah River Laboratory Report DP – 668 (1962).
3. Lᴏɴɢ, E., *et al.* Detection of faulty fuel elements. *J. Brit. Nuc. Eng. Conf.* April (1957).
4. Hᴇᴀʟᴛʜ, R. L. Fission product monitoring in reactor coolant water. *Nucleonics*, Dec. (1957).
5. Mɪʟʟᴀʀ, C. H., *et al.* Delayed neutron monitors for the detection of sheath failures in reactor fuel rods. *Nucl. Sci. and Eng.* **2** (1957).

6. GORDON, C. M., and HOOVER, J. I. Burst slug Cerenkov detector. *Nucleonics*, Jan. (1957).

7. HARLEN, F. Gas monitoring equipment for Berkeley Power Station. *Nucl. Eng.* April (1958).

8. ALIGA-KELLEY, D. B.S.D. in water cooled reactors. *Nucl. Power*, July (1959).

9. IGGULDEN, D. Burst slug detection in the A.G.R. *Nucl. Power*, May (1961).

10. KAUFMAN, J. L. Fuel element leak detection. *Nuclear Safety* **5,** 1 (1963).

11. KRITY, W. R. An automatic gas chromatograph for monitoring reactor fuel failures, Parts I to IV. *U.S.A.E.C. Reports* DP 356 (1959), DP 379 (1959), DP 593 (1961), DP 668 (1962).

12. MEGY, J. and ROGUIN, A. Detection of ruptures of the cans in French power reactors. *C.E.A. Report* 2051 (1961).

13. DENT, K. H. and GROSSMITH, G. W. Uranium fuel handling. *J. Brit. Nuc. Eng. Conf.* April (1951).

14. RODGER, W. A. Safety problems associated with the disposal of radioactive waste. *Nuclear Safety* **5,** 4 (1964).

15. KLEBER, I. A. Design requirements for nuclear reactor handling equipment. *Trans. Am. Nucl. Soc.* Sec. 10.4, Vol. 1 (1958).

CHAPTER 9

Operation at Power

9.1. INTRODUCTION

Start-up procedures have been discussed in Chapter 4. The initial start-up of a reactor is carried out on the satisfactory completion of the reactor physics experiments, control rod calibrations and the fuel element leak detection gear tests (described in Chapters 6, 7 and 8). The initial raising to power which is known as the power phase of commissioning, follows a similar procedure to that of a normal start-up with the exception that several additional balances at particular power levels are included. Measurements are made at each balance position in order to ensure that the procedure is carried out with the maximum safety and to check that no unusual reactor conditions have arisen. The results obtained at each stage are fully analysed before proceeding to the next balance position. On attaining full power for the first time measurements are made of every possible parameter to ensure that the anticipated conditions have been achieved.[1, 3, 9, 10, 11, 12, 13]

During the power phase of commissioning and on a routine basis throughout operation at power the following measurements are carried out:

(1) Temperature surveys and assessment of temperatures.
(2) Flux distribution measurements.
(3) Reactor heat balance.
(4) Determination of xenon poisoning concentration.

(5) Measurement of reactivity effects due to long-term changes.

(6) Reactivity balance.

(7) Temperature coefficient measurements.

(8) Control rod calibrations.

(9) Shield radiation surveys.

9.2. TEMPERATURE SURVEYS AND TEMPERATURE ASSESSMENT

The observation of local values of temperature is desirable in all reactors but its importance and the practicability of in-core measurements varies from reactor to reactor. In-core temperature measurements are often required for maximising the power performance of the core whereas in some reactors the value of the information obtaining may not be worth the expense of installation.

In the case of pressure-tube and individual coolant channel reactors it is usually easy to install adequate temperature and channel flow devices. Temperatures can usually be measured for individual channels using instruments located outside the core area. Some of the solid moderator reactor types are particularly favourable for in-core instrumentation and the measurement of outlet channel temperatures yields information for assessing the maximum temperature in the reactor and for increasing the power output by adjustment of the temperature distribution across the core. This topic will be dealt with in detail later.

In liquid moderated reactors of the pressure vessel type the core is immersed in a moderator-coolant and the difficulty of in-core measurement is greater and less use can be made of the information.

Discrete cooling channels are not well defined for each fuel element in many water reactors as the coolant is free to move from one element to another within an assembly of elements. Thus locally measured fuel element outlet temperatures do not yield useful information. In addition the thermal neutron flux gradients in water reactors are very large and so neighbouring fuel elements may operate at quite different power densities.

In spite of these difficulties in-core temperature measurement has been used to a large extent in large water moderated reactors. The information obtained is used to evaluate the position and magnitude of hot spots which are likely to occur. Local high-temperature regions must still be evaluated by calculation but experimental information yields a better assessment than would be obtained without reference to actual core temperatures. In view of the lengthy methods required for temperature assessment the degree to which the results are used to direct operations is limited. The difficulties of calculating temperature distributions in boiling water reactors are greater due to the interdependence of steam-void distribution, power distribution and reactivity and hence more use is made of in-core measurements in this case.

Complete temperature surveys involving the measurement of all reactor temperatures are carried out on a routine basis during reactor operation. Complete scans are also carried out following changes in reactor conditions and changes in the positioning of control or flattening absorbers. A large number of temperatures are measured in power reactors. For instance in a gas-cooled natural uranium reactor it is usual to measure of the order of 100 to 150 fuel can temperatures, 300 to 400 channel gas outlet temperatures, 200 to 300 graphite moderator temperatures and 50 to 100 pressure vessel and shield temperatures. In order to record and display these readings analogue and/or digital systems are provided using multi-point recorders and automatic data logging.

Obviously it is quite impossible for an operator to observe more than a few temperatures in the reactor. Certain selected thermocouples are therefore used to display temperature readings at the control desk for control purposes. It is essential for the operator to ensure that the maximum temperature allowed by the plant limitations is not exceeded. Consequently the choice of the fuel and gas outlet temperatures to be displayed is extremely important and is decided by carrying out a temperature assessment study as will be described later (Section 9.2.2).

It is also necessary to record temperatures as rapidly as possible since temperature changes can occur quickly. Automatic data loggers are provided for this purpose. During normal operation it is usual to select a number of representative thermocouple positions (of the order of 20 to 25% of the total number) to be scanned continuously and automatically. These thermocouples are scanned by a number of rotating switches and relay contacts. The voltages obtained are converted to digital form and these signals used to operate an automatic print-out system. The remaining thermocouples may also be displayed by a similar system but these are only recorded during normal operation at fixed intervals. Eight hours is considered to be an adequate period. The initiation of a complete temperature scan may be started by the operator at any time or devices may be incorporated so that the system automatically records a complete scan if unusual temperatures occur. This can be done by comparing the readings obtained from the continuously recorded thermocouples with pre-set standard voltages. If any difference occurs then a complete scan is initiated.

9.2.1. Temperature Limitations

Temperature limitations are imposed on reactor core temperatures in order to ensure that there is a minimal risk of damage to the fuel elements or a reactor fire. The power output of the reactor is dependent upon the allowed operating temperatures in the core. Therefore from an efficiency point of view it is important to operate with high fuel element temperatures. The criteria used to determine the temperature limitations must be chosen as realistically as possible otherwise severe restrictions will be placed on the overall economy of the plant.

A maximum allowed fuel element temperature is chosen to ensure that in the event of the maximum credible accident occurring a specified probability for any fuel element catching fire (or melting) is not exceeded. The probability of a fire occurring is calculated from temperature distributions in the reactor and involves determining the uncertainties in the mean and standard deviations in the temperature distributions.[2] The calculation of reactor probabilities from temperature measurements is a complex problem and assumptions are often made to ease the task. The types of probability calculations used vary but may be based on a treatment given by Stevens.[4]

The magnitude of the maximum allowed temperature is calculated assuming that a probability of a fire is not greater than 1 in 100 or 1 in 1000 dependent upon the inherent stability and safety of the particular reactor. A probability of 1 in 100 is considered a reasonably safe margin in gas-cooled reactors. The probability calculation is dependent upon the ignition temperature of the fuel canning material and on the maximum value of the temperature transient which occurs as a result of the maximum credible accident.

Ideally in order to achieve maximum possible power output

the actual measured maximum temperatures should equa
the limiting value determined by the probability calculations
In other words the maximum indicated fuel temperature shoulc
equal the maximum allowed temperature. However, this i
not possible as margins must be allowed to ensure that the
maximum allowed temperature is not exceeded at any poin
in the reactor. Only a limited number of fuel temperatures are
measured. Therefore a detailed analysis and assessment mus
be carried out to determine the maximum value of the indi
cated temperature. This involves taking into account the
possible systematic and random variations of fuel tempera
ture.

9.2.2. Temperature Assessment

A temperature assessment is carried out using the results o
a complete temperature scan to ensure that temperatures withi
the reactor are below the permissible limits, and to provid
operating values for the indicated temperatures displayed a
the control desk. The temperature assessment will ensure tha
maximum possible power is attained within the plant limi
tations. It is essential to be able to carry out a reasonabl
straightforward analysis which yields results quickly. Compli
cated assessments taking long periods of time to carry ou
are of little practical use to plant operators. A variety o
methods are available and it is proposed to outline the mai
features of a typical method.

Three temperatures are specified in a temperature assess
ment. These are:

(a) *Maximum allowed temperature*

This is the maximum permissible level which must never b
exceeded and which is determined by the plant limitation
(Section 9.2.1).

(b) *Maximum assessed temperature*

This is the estimated maximum fuel temperature at any point in the reactor obtained as a result of an assessment of all measured temperatures.

(c) *Maximum indicated temperature*

This is the particular temperature which is displayed to the operator at the control desk. It is chosen from the available measured temperatures and is used by the operator for control to ensure that the plant is operated within the specified limits. It is not necessarily the maximum measured temperature as will be seen later.

The following measurements are made prior to commencing the assessment:

(1) Fuel element temperatures.

Thermocouples are usually attached to one particular fuel element in a channel. In addition a number of channels may be instrumented with thermocouples attached to several fuel elements in the channel to obtain axial temperature variation.

(2) Channel gas outlet temperatures.
(3) Radial and axial flux distributions.
(4) Coolant flow in each fuel channel (usually obtained from flow measurements during the commissioning tests).
(5) Reactor heat balance.

The determination of the maximum assessed temperature and maximum allowed temperature is then carried out as follows:

1. Plot channel gas outlet temperatures (T_o) against core radius.

The systematic variations may be eliminated by drawing a smooth curve through the plotted points. The smooth curve is drawn to represent the trend of T_o rather than the arithmetical average at a particular radius. The curve is always biased to higher temperature values when in doubt so that the errors produce an overassessment and thereby err on the side of safety.

2. Determine the variation of channel heat rating with radius.

    If T_o and T_i are the channel outlet and inlet temperatures and F_c the channel flow, then the channel rating R is given by

    $$R = KF_c(T_o - T_i). \qquad (9.1)$$

    The constant of proportionality K in equation (9.1) may be determined by equating the total heat output of the reactor to the sum of the channel ratings over the reactor. This may be evaluated graphically using the values of T_o from the T_o-radius distribution and the values of F_c at a particular radius from the commissioning measurements.

    Substituting in equation (9.1) then the variation of channel rating R with radius may be determined.

3. Plot the fuel element temperatures for elements at a specified axial position T_E against radius.

    A smooth curve is again drawn through the points to eliminate systematic variations.

4. Plot the axial temperature variation of T_E at different radii using the results from any instrumented channels or from the theoretical estimated axial temperature variation.

5. Construct the maximum fuel element temperature (T_M) distribution with radius from the results of stages 3 and

4, and bias the plot to values of high temperature as before.

The values of T_M may be the same as T_E dependent on the positioning of the fuel element thermocouples.

6. The distribution of T_M against radius will enable the maximum assessed temperature to be deduced.

7. The random variation in fuel element temperature will be determined to estimate the expected uncertainty in the maximum assessed temperature and in the T_M–radius distribution.

Random variations arise due to differences in the heat transfer properties, minor channel to channel flow variations and errors in the measuring instrumentation. These variations may be estimated by considering the results obtained from groups of instrumented channels situated symmetrically with respect to the centre of the reactor. For each group the average temperature and the algebraic difference of each member from the average is found. Hence an estimate may be made of the standard deviation of random fuel element temperature variations.

8. The maximum assessed temperature is then taken as the maximum value of T_M plus the maximum expected error due to the random variations.

From this assessment the maximum allowed temperature for a given set of reactor conditions is also calculated. The radial variation of the maximum temperature transient occurring as a result of the maximum credible accident is calculated using the channel rating distribution (item 2). Applying the deduced random temperature variation, the probability of a maximum transient temperature occurring in the reactor which is greater than the ignition temperature may be calculated. This enables the maximum allowed temperature to be specified for the particular reactor conditions at which the survey was carried out.

Thus using this treatment values of the maximum allowed temperature and the maximum assessed temperature in the reactor may be obtained for a given set of conditions.

A particular fuel element temperature (or a small number of fuel element temperatures) must be chosen from the measured values to be displayed at the control desk. This will be used to limit the temperatures in the whole reactor and is the maximum indicated temperature.

The particular fuel element to be used for this purpose must be chosen from one of the hottest measured fuel elements but must be in a region which is representative of reactor conditions. In other words it would not be realistic if a very hot fuel element were chosen in an abnormal region. In this case the measured temperature variations would not be representative of the general trend throughout the reactor. Thus it is more important to obtain true representation than merely to choose the hottest fuel element.

The maximum value of the chosen indicated temperature is determined from the temperature survey and assessment. If it is found that the maximum assessed temperature is equal to the maximum allowed temperature (i.e. already at the limiting value) then the maximum indicated temperature is the value measured during the temperature survey. This value must not be exceeded. If the maximum assessed temperature is found to be less than the limiting value then the value of the maximum indicated temperature is given by the expression

$$T_{\text{ind}} = T_E + \Delta T_M \frac{(T_E - T_i)}{(T_M - T_i)} \tag{9.2}$$

where T_E is the fuel temperature of the chosen fuel element,

T_i is the coolant inlet temperature,
T_M is the maximum assessed temperature, and
ΔT_M is the temperature difference between the maximum assessed and maximum allowed temperatures.

It is common practice to determine the maximum value for a number of possible indicated fuel element temperatures.

As a result of the temperature survey and assessment the operator will be given a number of fuel element temperatures to be recorded at the control desk and the maximum values which must not be exceeded.

9.3. OPERATIONAL FLUX PLOTS

Provision is made in reactors to enable neutron flux distributions to be measured during operation at power.[5, 6, 7] Basically this is carried out by lowering a wire into the reactor through the plugged holes in the top shield normally used for fuel charge and discharge or in specially installed thimbles. The wire is irradiated for a suitable time, removed from the core and the induced activity of the wire determined. In some cases arrangements are made to detect the activity of the wire as it is withdrawn from the core.

The wire used for this purpose must have suitable radioactive properties (i.e. neutron absorption cross-section and half-life) and suitable mechanical properties for insertion and withdrawal into the core. It is found that stranded tungsten wire is the most convenient material. The structural properties are ideal for the purpose and the half-life of the radioactive decay is 24 hr.

The tungsten wire is wound on a drum in a special winch unit which can be attached to the top of a hole used for charge and discharge purposes. The shield plugs in the charge holes are removed and replaced with special flux scanning plugs using the charge/discharge machinery. The winch unit is placed on top of the plugs and a hoist mechanism lowers the tungsten wire into the core. A typical arrangement is shown in Fig. 9.1.

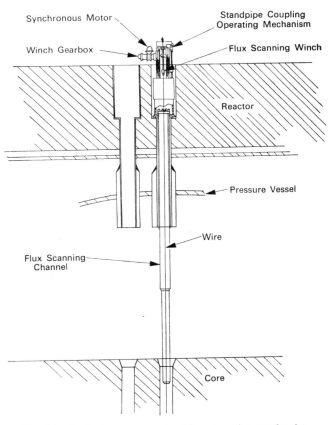

FIG. 9.1. Typical arrangement of flux scanning mechanism

The hoist mechanism and winch unit are enclosed in a steel vessel surrounded by lead shielding to provide protection when the irradiated wire is wound into the winch. The apparatus is designed so that access to the reactor vessel is possible with the reactor at full power and at full operational pressure. A series of valves, rotating shields, purging units and temporary shielding is provided to ensure maximum safety during the operations.

The wire is lowered into the reactor core and irradiated for a sufficient period to give adequate induced activity for measurement purposes. Using tungsten wire of diameter 0·1 in. in neutron fluxes of the order of 10^{13} n/cm^2 sec then an induced activity of approximately 1 mc per cm of wire can be expected for an irradiation period of 30 min. This level of activity is sufficient to produce measurable ionisation currents in a sensitive ionisation chamber.

The insertion and withdrawal speeds for the wire must be short in comparison with the irradiation time. This is necessary in order to avoid significant additional irradiation of the end of the wire as it passes through the reactor core. The hoists provide speeds of the order of 6 in./sec and hence 40 ft of wire will be inserted in a period of 80 sec.

At the end of the irradiation the wire is withdrawn into the winch unit. This unit is removed from the reactor and positioned above an instrumentation well. The wire is lowered into the well and is passed through a hollow ionisation chamber and the induced activity of the tungsten wire is measured. The ionisation current is fed through a D.C. amplifier and used to drive a potentiometric pen recorder. The recorder chart is moved in synchronisation with the passage of the wire and thus provides a direct graphical record of the activity along the wire. The ionisation chamber is usually shielded to prevent errors arising as a result of radiation from the wire wound on the winch. A typical hollow ionisation chamber suitable for flux scanning equipment has an active length of 10 cm and an activity of 1 mc/cm over this length will produce a current of approximately 10^{-9} A. The wire is wound through the ion chamber at a speed of $\frac{1}{2}$ in./sec. As the induced activity is proportional to the neutron flux the recorder will represent the axial flux distribution at a particular position in the reactor.

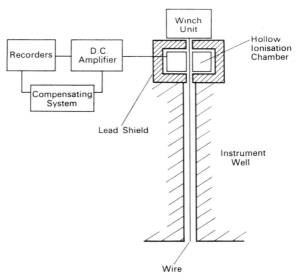

FIG. 9.2. Method of determining activity on a flux scanning wire

Figure 9.2 indicates schematically the method of determining the activity on a flux scanning wire.

It has been pointed out that errors can be introduced due to the fact that the lower part of the wire receives additional irradiation during the insertion and withdrawal process. In some cases compensating networks are incorporated to correct for this effect by varying the sensitivity of the D.C. amplifier associated with the ion chamber as the wire moves through the chamber.

In order to obtain a complete flux scan of the reactor the procedure is repeated and axial flux distributions obtained at several radial positions. Radial flux distributions are obtained at different heights from the axial measurements.

This method of flux distribution measurement suffers from a number of disadvantages. There is a limit to the number of wires which can be irradiated at any one time due to the expense and inconvenience of providing large numbers of instrumentation facilities. Two positions are normally provided which means that only two wires can be measured at any one time.

In practice it is found that to obtain axial flux measurements at 20 different positions in the reactor a total period of approximately 20 hr is required. Consequently as wires are being irradiated in the core at different times it is essential to maintain reactor conditions constant throughout this period in order to correlate the measurements. This introduces the main source of error into the method as it is often quite difficult to avoid changes in operating conditions during the flux scanning period. Flux scans are not often used for detailed flux shaping experiments. In this case the measurement of fuel and channel outlet temperatures are a much better guide to flux conditions in the core.

9.4. REACTOR HEAT BALANCE

Reactor heat balances are carried out on a routine basis in order to check the accuracy of the nuclear power recording instruments. Reactor power is measured by determining the thermal neutron flux near the core or in an external thermal column using ionisation chambers (see Chapter 3). The ionisation chamber current is amplified and used to operate indicators and recorders calibrated in terms of the amount of heat generated in the reactor core.

During the initial loading of fuel and the reactor physics commissioning experiments the low power instrumentation is calibrated by foil activation techniques (see Chapter 6), the heat generated in the core in this case being deduced from flux distribution measurements. These methods are not easy to carry out during operation at power and are not sufficiently accurate for reliable estimates of the reactor power. Consequently at power the nuclear instruments are calibrated by carrying out heat balances. The heat output of the reactor is determined at steady conditions and the nuclear instruments adjusted by alteration of the recorder scales or by physical movement of the ionisation chambers to give a true indication of the actual power.

The following procedure is adopted to determine a heat balance:

1. Allow reactor conditions to stabilise.
2. Record the power measurements from all the nuclear instruments.
3. Determine the average reactor outlet temperature (T_o) and outlet coolant pressure (P_o).
4. Measure the apparent coolant flow.

    The flow measurement is usually only correct absolutely at a particular set of conditions (i.e. at particular values of temperature (T) and pressure (P)).
5. Determine the absolute coolant flow.

    A density correction factor must be applied to the apparant flow which is dependent upon the relevant temperatures and pressures. When an orifice plate is used for flow measurement then

$$\text{absolute flow} = \text{apparent flow} \sqrt{\frac{T}{P} \cdot \frac{P_o}{T_o}}$$

6. Determine the average reactor inlet temperature (T_i).
7. The heat content of the coolant at the reactor inlet and outlet is then determined from the values of T_o, T_i, the coolant flow, and the specific heat of the coolant.
8. The reactor heat output is the change in heat content between the reactor inlet and outlet.

On large power reactors with a number of heat exchangers the inlet and outlet heat content of the coolant for each heat exchanger circuit is determined. The total heat output must include any additional heat supplied to the reactor from external sources such as the gas circulators. The heat generated in each circulator is measured and an allowance made for this effect.

The reactor output may also be measured from heat balances obtained from the steam generation conditions at the heat exchangers on a power station. These provide an additional check on overall plant conditions but are not as accurate as a reactor heat balance due to heat losses in the circuits.

9.5. XENON POISONING

A very important fission product as far as reactor operation is concerned is the isotope xenon-135 on account of its large thermal neutron absorption cross-section ($3 \cdot 5 \times 10^6$ barns). The presence of xenon-135 in a reactor will reduce the k_{eff} of the system considerably and additional excess reactivity must be available to overcome this effect.

Xenon is formed as a direct fission product with a fission yield of $0 \cdot 3\%$ and as a result of the decay of the fission product tellurium-135 with a fission yield of 5·6%. The decay of tellurium produces the main proportion of xenon in an operating reactor in accordance with the decay scheme

$$\text{Te-135} \xrightarrow[\text{2 min}]{\beta^-} \text{I-135} \xrightarrow[\text{6·7 hr}]{\beta^-} \text{Xe-135} \xrightarrow[\text{9·2 hr}]{\beta^-} \text{Cs-135} \xrightarrow[\text{2}\times\text{10}^4\text{ yr}]{\beta^-} \text{Ba-135} \quad \text{(Stable)}$$

Tellurium decays with a half-life of 2 min to iodine-135 which is not a strong neutron absorber. This decays with a half-life of 6·7 hr to produce xenon-135 which itself decays with a half-life of 9·2 hr to form a non-neutron absorber caesium-135.

In order to calculate the amount of xenon present in a reactor it may be assumed that iodine-135 is formed directly as a result of fission and only the iodine to caesium part of the decay chain will be considered. Assuming that iodine-135 is formed directly with a fractional yield γ_I of 5·6%; then

Rate of formation of I-135 = $\gamma_I \Sigma_f \phi$ per cm³ per sec

where Σ_f is the macroscopic fission cross-section, and
ϕ is the thermal neutron flux.

Rate of radioactive decay of I-135 = $\lambda_I I$ per cm³ per sec

where I is the number of nuclei present per cm³, and
λ_I is the decay constant.

Rate of decay of I-135 by neutron capture = $\sigma_I I \phi$ per cm³/sec

where σ_I is the neutron absorption cross-section.

Hence rate of increase of I-135 nuclei will be given by

$$\frac{dI}{dt} = \gamma_I \Sigma_f \phi - I(\lambda_I + \sigma_I \phi) \tag{9.3}$$

At equilibrium the number of I-135 nuclei will be given by

$$I_0 = \frac{\gamma_I \Sigma_f \phi}{\lambda_I + \sigma_I \phi} \tag{9.4}$$

The concentration of I-135 at any time t may be obtained by integrating equation (9.3) (Ref. 3, Chapter 4) to give

$$I(t) = I_0[1 - \exp-(\lambda_I + \sigma_I\phi)t] + I(0) \exp-(\lambda_I + \sigma_I\phi)t \quad (9.5)$$

where $I(0)$ is the initial number of iodine-135 nuclei.

If the time period commences at reactor start-up then the initial number of nuclei $I(0)$ will be zero and so in this case

$$I(t) = I_0[1 - \exp-(\lambda_I + \sigma_I\phi)t] \quad (9.6)$$

The absorption cross-section of iodine-135 σ_I is small and therefore $\sigma_I\phi$ may be neglected in comparison with λ_I. So the expressions for the equilibrium concentration of iodine-135 and the concentration at time t reduce to

$$I_0 = \frac{\gamma_I \Sigma_f \phi}{\lambda_I} \quad (9.7)$$

and

$$I(t) = I_0[1 - \exp-\lambda_I t] \quad (9.8)$$

Similarly the increase in the number of nuclei of the isotope xenon-135 may be derived. We have that

Rate of formation of Xe-135 by fission $= \gamma_X \Sigma_f \phi$ per cm³ per sec

Rate of formation of Xe-135 by decay of I-135 $= \lambda_I I$ per cm³ per sec

Rate of radioactive decay of Xe-135 $= \lambda_X X$ per cm³ per sec, and

Rate of decay of Xe-135 by neutron capture $= \sigma_X X\phi$ per cm³ per sec.

Hence

$$\frac{dX}{dt} = \gamma_X \Sigma_f \phi + \lambda_I I - X(\lambda_X + \sigma_X\phi) \quad (9.9)$$

The equilibrium concentration is given by

$$X_0 = \frac{\lambda_I I_0 + \gamma_X \Sigma_f \phi}{\lambda_X + \sigma_X\phi} \quad (9.10)$$

or from equation (9.7)

$$X_0 = \frac{(\gamma_I + \gamma_X)\Sigma_f \phi}{\lambda_X + \sigma_X \phi} \qquad (9.11)$$

By integration of equation (9.9) and taking the time period from the commencement of reactor start-up (i.e. assuming that the initial concentrations of iodine and xenon are zero), an expression for the concentration of xenon-135 at any time t is obtained. This is

$$X_t = X_0[1 - \exp -(\lambda_X + \sigma_X \phi)t]$$
$$+ \frac{\lambda_I I_0}{(\lambda_X + \sigma_X \phi - \lambda_I)} [\exp -(\lambda_X + \sigma_X \phi)t - \exp -\lambda_I t] \quad (9.12)$$

The poisoning of a reactor by an absorber is defined as the ratio of the number of thermal neutrons absorbed by the absorber to the number absorbed by the fuel,

i.e.   poisoning, $P = \dfrac{\Sigma_a}{\Sigma_u}$ (9.13)

where Σ_a and Σ_u are the macroscopic absorption cross-sections for the absorber and the fuel. (The macroscopic absorption cross-section is related to the microscopic cross-section by the expression $\Sigma = N_\sigma$ where N is the number of nuclei per c.c.)

The relationship between the poisoning, P, and reactivity may be derived by considering the effect of the absorber (poison) on the thermal utilisation and the k_{eff} of the reactor.

If it is assumed that the reactor is just critical without the poison then the reactivity change due to the poison will be given by

$$\varrho = -\frac{k_{ea} - 1}{k_{ea}} \qquad (9.14)$$

where k_{ea} is the effective multiplication factor of the reactor with the poison added.

The thermal utilisation with poison (f_a) and without the poison (f) are given by

$$f = \frac{x}{x+1}$$

and

$$f_a = \frac{x}{x+1+\dfrac{\Sigma_a}{\Sigma_m}} = \frac{x}{x+1+P_x}$$

where Σ_m is the macroscopic absorption cross-section of the moderator and x is the ratio Σ_u/Σ_m.

Then from equation (9.14) the negative reactivity due to the addition of the poison is given by

$$\varrho = \frac{f_a - f}{f_a} = -\frac{P_x}{x+1} \tag{9.15}$$

If the ratio of Σ_u/Σ_m is much greater than 1 then the value of the poisoning is equal to the negative reactivity produced.

The value of the poisoning at any time t and at equilibrium is obtained by using equations (9.13), (9.12) and (9.11), i.e.

$$P_t = \frac{X_t \sigma_X}{\Sigma_u} \tag{9.16}$$

and

$$P_0 = \frac{(\gamma_I + \gamma_X)\Sigma_f \sigma_X \phi}{(\lambda_X + \sigma_X \phi)\Sigma_u} \tag{9.17}$$

The equilibrium value for xenon poisoning P_0 in a reactor is dependent on the value of the neutron flux as can be seen from equation (9.17). Typical values of P_0 for an uranium-235 reactor are plotted in Fig. 9.3 against neutron flux.

For fluxes less than 10^{11} n/cm² sec the $\sigma_X \phi$ term in the denominator of equation (9.17) may be neglected and the value of P_0 depends on the relative magnitudes of the appropriate cross-sections. It is found that the value of P_0 is negligible below a flux of 10^{11} and small in the range 10^{11} to 10^{12}.

The magnitude of the poisoning gradually increases with neutron flux. For very high fluxes λ_X in the denominator of equation (9.17) can be neglected in comparison with $\sigma_X \phi$ and hence a limiting value of P_0 is reached. In practice this limiting value is attained for fluxes in excess of 10^{15} n/cm² sec and is of the order of 5% in reactivity.

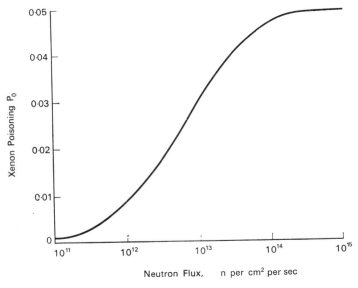

Fig. 9.3. Values of equilibrium xenon poisoning

Regarding reactor operation it is found that as the xenon poison builds up to its equilibrium value reactivity must be released to maintain a steady power level. This is done by withdrawing the control rods in as uniform a pattern as possible to maintain constant power and flux levels in the reactor. A decrease in reactivity is produced as xenon builds up and eventually an equilibrium level is achieved. An important aspect of xenon poisoning is the change which occurs following a reactor shut-down. As has already been pointed out the half-life for the decay of iodine-135 (6·7 hr) is shorter than the half-life for the decay of xenon-135 (9·2 hr). Consequently on shut-down the concentration of xenon in the reactor will increase above the equilibrium value as it is formed by the decay of iodine faster than it decays. The xenon poisoning will increase therefore on shut-down up to a maximum dependent upon the decay of iodine-135 and then gradually decay away completely dependent upon its own half-life.

Let us consider the case when a reactor has operated at power for a sufficient period to allow the equilibrium concentrations of iodine and xenon to be achieved. On shut-down the neutron flux is immediately reduced to zero and the variation of iodine-135 with time following the shut-down will be given by putting $\phi = 0$ in equation (9.5),

i.e.
$$I(t) = I_0 \exp - \lambda_I t \qquad (9.18)$$

In this case the initial concentration of iodine is the equilibrium value I_0. Similarly to determine the variation of xenon with time after shut-down equation (9.9) is integrated, and putting $\phi = 0$ and the initial concentration of xenon equal to the equilibrium value X_0 we get

$$X(t) = \frac{\lambda_I}{\lambda_X - \lambda_I} I_0 [\exp - \lambda_I t - \exp - \lambda_X t] + X_0 \exp - \lambda_X t \quad (9.19)$$

The values of I_0 and X_0 are dependent on the equilibrium value of the neutron flux and therefore the build-up of xenon following a shut-down is dependent upon the initial operational flux level. The variation of the xenon poisoning with time following a shut-down is obtained by substituting the value of $X(t)$ from equation (9.19) into (9.16).

The time required to reach the maximum xenon concentration after shut-down is an important period from an operational aspect. It is obtained by differentiating equation (9.19) with respect to time and equating the result to zero. Hence we get

$$t_{max} = \frac{1}{\lambda_X - \lambda_I} \log_e \frac{\lambda_X}{\lambda_I} \left[1 - \frac{\lambda_X - \lambda_I}{\lambda_I} \cdot \frac{X_0}{I_0} \right] \qquad (9.20)$$

The increase in xenon concentration following reactor shut-down can be considerable. In the case of a typical gas-cooled graphite-moderated power reactor, an equilibrium xenon concentration equivalent to a negative reactivity of the order of 2·0% is achieved. This occurs with a maximum operating flux level of the order of 2×10^{13} n/cm² sec. On shut-down from this flux level the negative reactivity worth due to xenon increases to a peak value of 2·6% in a period of approximately 7 hr. The variation of negative reactivity with time is shown in Fig. 9.4. Obviously in order to start up within 7 hr of shut-down an additional 0·6% in reactivity must be available. This problem is not a serious one as in any case total time required to start up a power reactor of this type is considerably greater than 7 hr and the xenon will therefore have decayed.

In the case of high flux research reactors the problem of the increase of xenon concentration after shut-down introduces a severe limit on reactor operation. For a flux of 2×10^{14} n/cm² sec the equilibrium value of the negative reactivity due to xenon is of the order of 4·6%. On shut-down this will rise to a peak value of the order of 50% in reactivity in a period of 11 hr. Hence it is quite out of the question to build in additional reactivity to enable start-up to be achieved during the xenon build-up. Start-up is therefore impossible until the xenon has decayed. The time to reach the equilibrium level in this case is of the order of 60 to 70 hr, hence imposing a limitation on the operation of the reactor. The problem can be avoided by shutting the reactor down in a controlled manner so that the flux is reduced gradually. In high flux research reactors additional reactivity is built in to enable start-up to recommence within a period of 20 to 30 min of shut-down. In this case the total negative reactivity due to xenon will be approximately 7%.

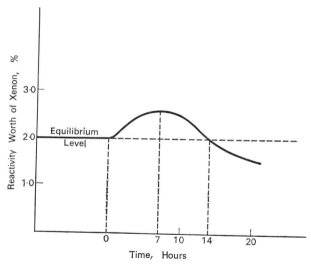

Fig. 9.4. Variation of reactivity absorbed by xenon poisoning with time after shut-down (equilibrium flux level 2×10^{13} n/cm² sec).

9.5.1. The Measurement of the Reactivity Worth of Xenon Poisoning

A measurement of the change in reactivity due to the presence of xenon in a reactor core is required during operation.[8] The reactor is first balanced at a particular power level. This balance is then maintained for a period of 48 hr by movement of the control rods only. The rods should be moved in such a way that the core temperatures and flux distributions are maintained as much as possible. The control rods are withdrawn during the period of the experiment to release reactivity to balance that absorbed as xenon builds up in the reactor. The reactivity worth of the xenon poisoning is obtained from the rod movement required to maintain a balance and the previously measured control rod calibrations. A number of corrections may have to be made to allow for any differences in neutron flux shapes in the core which occur during the experiment. However these effects can be reduced considerably by the suitable choice of rods to be moved and their distribution.

9.6. REACTIVITY EFFECTS DUE TO LONG-TERM CHANGES

9.6.1. Samarium Poisoning

The isotope neodymium-149 is produced in reactors as a result of fission with a yield of approximately 1·4%. This decays with a half-life of 1·7 hr to promethium-149 which in turn decays to samarium-149 with a half-life of 47 hr. The isotope samarium-149 is a stable nuclide and is important from an operational aspect due to its high thermal neutron absorption cross-section ($5·3 \times 10^4$ barns). The production of samarium-149 will therefore lead to a reduction in reactivity during operation in a similar way to xenon. Let us consider the promethium to samarium part of the decay chain and assume that promethium is produced directly as a result of fission. This is a reasonable assumption as the half-life of neodymium is short in comparison with that of promethium.

The same treatment as used previously (Section 9.5) may be applied, the main difference being that samarium-149 stable and does not decay.

The build-up of samarium-149 with time is identical to that for xenon given by equation (9.9) with the exception that the decay constant for samarium is zero. Hence, the build-up may be expressed as

$$\frac{dS}{dt} = \gamma_S \Sigma_f \phi + \lambda_P P - S \sigma_S \phi \qquad (9.21)$$

where S and P refer to samarium and promethium.

The equilibrium concentration is given by an expression similar to equation (9.10), i.e.

$$S_0 = \frac{\lambda_P P_0 + \gamma_S \Sigma_f \phi}{\sigma_S \phi} \qquad (9.22)$$

where P_0 the equilibrium concentration of promethium-149 is given by

$$P_0 = \frac{\gamma_P \Sigma_f \phi}{\lambda_P}$$

assuming that the thermal absorption cross-section of promethium is negligible; and so by substitution into equation (9.22) we have that

$$S_0 = \frac{(\gamma_P + \gamma_S) \Sigma_f}{\sigma_S} \qquad (9.23)$$

The variation of the number of samarium-149 nuclei with time is similar to equation (9.12), i.e.

$$S_t = S_0[1 - \exp - \sigma_S \phi t] + \frac{\lambda_P P_0}{\sigma_S \phi - \lambda_P} [\exp - \sigma_S \phi t - \exp \lambda_P t] \quad (9.24)$$

It can be seen from equation (9.23) that the equilibrium value of the samarium poisoning is independent of the neutron flux in the reactor. It is found that for a uranium-235 reactor the equilibrium value is approximately $-1\cdot2\%$ in reactivity. However, it must be pointed out that for fluxes of the order of 10^{11} to 10^{12} n/cm^2 sec it requires a period of four years for the equilibrium concentration to be attained.

On shut-down the concentration of samarium-149 will increase to an asymptotic level which is dependent only on the decay of promethium. The variation with time from shut-down is given by an expression similar to equation (9.19) and is

$$S_t = S_0 + P_0(1 - \exp - \lambda_p t) \qquad (9.25)$$

The value of S_t is dependent on the neutron flux before shut-down. For a flux of 2×10^{14} n/cm^2 sec the asymptotic worth due to samarium build-up on shut-down is of the order of 4.0% in reactivity. The effect of samarium-149 poisoning during day-to-day operation of reactors does not present any undue problems.

9.6.2. Long-term Reactivity Effects

Other reactivity changes occur in operating reactors due to the build-up of poisons and due to the formation of fissile nuclei. In uranium reactors these changes result from

(a) Depletion of uranium-235.

 This produces a negative reactivity change due to the formation of neutron-absorbing fission products. The number of fission products increases with irradiation of the fuel.

(b) Build-up of plutonium-239, 240 and 241.

 As irradiation proceeds, uranium-238 nuclei capture neutrons and produce plutonium-239 by electron emission. This is a fissile material and hence a positive reactivity change occurs which gradually increases with the build-up of plutonium-239. Plutonium-239 is eventually depleted by the formation of neutron-absorbing fission products.

Plutonium-239 also captures neutrons and produces plutonium-240 which is non-fissile and a neutron absorber. After several 1000 MWD of irradiation a certain amount of plutonium-241 will be produced which is fissile and will tend to increase the reactivity of the core. In most cases it is not usually necessary to consider the formation of plutonium-241.

Thus as irradiation of the fuel proceeds a number of reactivity changes occur as a result of the fuel depletion and the build-up of plutonium-239. These long-term reactivity changes are determined by carrying out a reactivity balance (see Section 9.7) at frequent intervals during operation of the reactor at power. From the reactivity balance the total excess reactivity of the core may be deduced. The variation of the excess reactivity with irradiation is a measure of the long-term changes.

On balancing a uranium-235 reactor at power there is initially a decrease in reactivity due to the build-up of xenon-135 to its equilibrium concentration. This takes place in a period

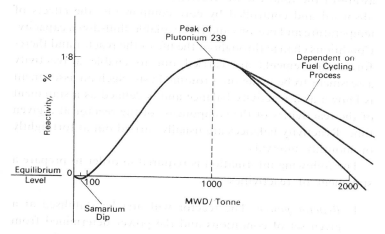

FIG. 9.5. Reactivity variation due to long-term changes

of 2 to 3 days. Following this the reactivity decreases due to the build-up of samarium-149 and other poisons, and then increases as the fissile plutonium-239 is produced. Eventually a peak is reached in the reactivity curve. After this further irradiation will gradually reduce the reactivity until eventually the reactor must be refuelled. In some cases fuel cycling is employed so that depleted fuel is replaced with new fuel on a continuous cycle. In this case the replacement of fuel will enable a reactivity equilibrium to be achieved in the reactor. Figure 9.5 indicates the variation of Δk_{eff} of the core with irradiation for a natural uranium reactor operating with a central neutron flux of 2×10^{13} n/cm^2 sec. The irradiation period commences when xenon equilibrium has been achieved.

9.7. REACTIVITY BALANCE

It is essential at any stage during operation at power to be aware of the total excess reactivity of the core, the reactivity absorbed and controlled by core components, the effects of temperature and poisons and the available shut-down capacity. Conditions change throughout the life of the reactor and therefore measurements are carried out to enable a reactivity assessment to be made on a routine basis. Such an assessment is known as a reactivity balance and is defined as a statement of the reactivities of the components of the reactor at a given time. Reactivity balances are usually carried out at fortnightly or monthly intervals.

The following information is required in order to prepare a statement of reactivities:

1. *Reactor power*. The reactor will first be stabilised at a given set of conditions and the power determined from a heat balance measurement.

2. *Flux distribution.* A complete flux scan will be carried out at uniform conditions.

3. *Temperature survey.* A complete temperature survey and assessment will be made.

4. *Built-in excess reactivity.* A value for the built-in excess reactivity of the unpoisoned core at 20°C will be estimated. This value is obtained from the initial commissioning measurements.

5. *Total reactivity absorbed.* The reactivity absorbed by partially or fully inserted control rods and other absorbers (i.e. flattening rods, samples, thermocouples, etc.) will be determined from initial commissioning tests and from the control rod calibrations.

6. *Reactivity effect of xenon.* Determined from control rod movement during the build-up of the xenon concentration.

7. *Long-term reactivity changes.* Obtained from previous reactivity balances and movement of control rods.

8. *Reactivity effects due to temperature changes.* The graphite and uranium temperature effects are obtained from the latest available measured values of the graphite and uranium temperature coefficients (see Section 9.8) and the weighted mean graphite and uranium temperatures. These values are determined from the temperature surveys and flux distribution measurements.

The weighted mean graphite temperature is defined as

$$\bar{T}_G = \frac{\int_0^{R_e} \int_0^{H_e} \phi^2 R T_G \, dR.dH}{\int_0^{R_e} \int_0^{H_e} \phi^2 R \, dR.dH} \tag{9.26}$$

and the weighted mean uranium temperature as

$$T_u = \frac{\displaystyle\int_0^{R_e} \int_0^{H_e} \phi^2 R T_u \, dR.dH}{\displaystyle\int_0^{R_e} \int_0^{H_e} \phi^2 R \, dR.dH} \qquad (9.27)$$

where ϕ is the neutron flux,

R_e the extrapolated radius and R the radius,

H_e the extrapolated height and H the height,

T_G the graphite temperature,

T_u the mean uranium temperature.

(a) Determination of the weighted mean graphite temperature, T_G

1. Divide the reactor into a number of axial and radial zones.
2. Plot T_G against height at the centre of each radial zone.
3. Plot T_G against radius (R) at the centre of each axial zone.
4. Determine ϕ^2 (from flux plots) and plot ϕ^2 against R.
5. Obtain values of $\phi^2 R$ and $\phi^2 R T_G$ at values of R corresponding to the centre of each axial zone and plot against R.
6. Integrate under the curves and deduce values for

$$x_1 = \int_0^{R_e} \phi^2 R \, dR$$

and

$$x_2 = \int_0^{R_e} \phi^2 R T_G \, dR$$

7. Plot values of x_1 and x_2 against height (H). Integrate under the curves and obtain values for

$$\int_0^{H_e} x_1 \, dH \quad \text{and} \quad \int_0^{H_e} x_2 \, dH$$

8. Substitute in equation (9.26) and estimate T_G.

(b) Determination of the weighted mean uranium temperature, $\bar{T}_u$

The weighted mean uranium temperature is obtained by a similar analysis to that described above to deduce $\bar{T}_G$. The reactor is divided into radial and axial zones and values for $\phi^2 R$ and $\phi^2 R T_u$ are plotted against R for each zone. Values of $\int_0^{R_e} \phi^2 R \, dR$ and $\int_0^{R_e} \phi^2 R T_u \, dR$ are determined and plotted against H. These values are integrated and hence $\bar{T}_u$ determined.

The main problem in obtaining $\bar{T}_u$ is that of estimating a value for T_u the mean uranium temperature. Fuel element temperatures are obtained from thermocouples attached to the surface of the fuel can and a method is required of deducing the mean uranium temperature from these measurements. A theoretical value for the temperature difference between the uranium surface temperature and the can temperature is usually available and measured values are often obtained from design rig experiments. If this temperature difference is ΔT, then the uranium surface temperature, T_S, is given by

$$T_S = T_C + \Delta T$$

where T_C is the measured can temperature.

The average power developed in a particular fuel element is dependent upon the temperature difference between the mean uranium temperature and the surface temperature and the thermal conductivity of the uranium. Hence if the average power per fuel element is determined and the thermal conductivity and T_S are known, the mean uranium temperature, T_u, may be calculated.

The average power per fuel element is calculated in the following way:

1. From the flux distribution measurements plot ϕR against

R for each axial zone corresponding to the centre of the fuel elements.

2. Integrate under the curves and obtain $\int_0^{R_e} \phi R \, dR$ for each zone.

3. The average power per fuel element is given by

$$P_{av} = \frac{A \int_0^{R_e} \phi R \, dR}{N} \qquad (9.28)$$

where A is a constant of proportionality, and

N is the number of fuel elements in each zone.

The reactor power P (determined from heat balance) is given by

$$P = A\Sigma \int_0^{R_e} \phi R \, dR \qquad (9.29)$$

and so

$$P_{av} = \frac{P}{N} \cdot \frac{\int_0^{R_e} \phi R \, dR}{\Sigma \int_0^{R_e} \phi R \, dR} \qquad (9.30)$$

Hence P_{av} may be determined and so T_u and $\bar{T}_u$ deduced.

9.7.1. **Typical Reactivity Balance**

Typical results obtained for a graphite-moderated natural uranium reactor are given in Table 9.1 at zero irradiation and at the peak of plutonium build-up.

9.8. TEMPERATURE COEFFICIENTS

Temperature variations in operating reactors produce a change in reactivity.[3, 8] The operational characteristics are therefore dependent upon the magnitude of the reactivity

TABLE 9.1. TYPICAL REACTIVITY BALANCES

(Column A—zero irradiation: Column B—at peak of plutonium build-up)

Fixed reactivity	A (%)	B (%)	Variable reactivity	A (%)	B (%)
1. Built in at 20°C and un-poisoned	+4·50	+4·50	1. Uranium temperature effect $(\alpha_u \cdot \bar{T}_u)$	−0·80	−0·80
2. Samples	−0·02	−0·02	2. Graphite temperature effect $(\alpha_G \cdot \bar{T}_G)$	−1·00	+0·60
3. Thermo-couples	−0·02	−0·02	3. Xenon poisoning	−1·80	−1·80
4. Miscellaneous absorbers	−0·06	−0·06	4. Reactivity build-up	0	+2·00
			5. Control rods 5.1 Flattening rods	−0·80	−0·80
			5.2 Bulk control rods	0	−3·60
Totals	+4·40	+4·40	Totals	−4·40	−4·40

change and whether it increases or decreases the multiplication factor.

Reactivity varies with temperature on account of:

(a) the change in energy of the thermal neutrons which alters the relative effect of the various reaction cross-sections of the reactor materials; and

(b) the change in reactor size due to density changes.

These two different effects are sometimes known as the nuclear temperature coefficient and the density temperature coefficient respectively.[3]

For a large reactor the relationship between k_{eff} and k_{∞} is given by equation (6.2), i.e.

$$k_{\text{eff}} = \frac{k_{\infty}}{1 + \alpha^2 M_z^2 + \beta^2 M_r^2} \qquad (9.31)$$

Reactivity ϱ is given by $\dfrac{k_e - 1}{k_e}$ and hence

$$\varrho = \frac{k_{\infty} - 1}{k_{\infty}} - \frac{\alpha^2 M_z^2 + \beta^2 M_r^2}{k_{\infty}} \qquad (9.32)$$

Using a simpler approach (Ref. 3, p. 255) we get

$$\varrho = \frac{k_{\infty} - 1}{k_{\infty}} - \frac{B^2 M^2}{k_{\infty}} \qquad (9.33)$$

where B^2 is the geometrical buckling, and

M^2 the migration area equal to $L^2 + L_s^2$.

With constant physical size α^2 and β^2 will remain constant with temperature change whereas M_z^2, M_r^2 and k_{∞} will vary. The values of M_z^2 and M_r^2 are determined by the slowing-down lengths and the diffusion lengths in the moderating material.

The diffusion length, L, is related to the macroscopic absorption cross-section Σ_a by

$$L^2 \alpha \frac{1}{\Sigma_a} \qquad (9.34)$$

In addition, in the thermal energy region it is known that absorption cross-sections are proportional to the inverse of the neutron velocity. Therefore, as the neutron energy is proportional to the absolute temperature, T, the absorption cross-section will be inversely proportional to the square root of the temperature and so (9.34) becomes

$$L^2 \alpha (T)^{1/2} \tag{9.35}$$

assuming that the variation of the scattering cross-section with temperature is negligible. This may also be expressed as

$$L_2^2 = L_1^2 \left(\frac{T_2}{T_1}\right)^{1/2} \tag{9.36}$$

where the subscripts 1 and 2 refer to temperatures T_1 and T_2.

The variation of k_∞ with temperature depends upon the thermal utilisation factor and the resonance escape probability. However, if it is assumed that all the absorbing materials follow the $1/V$ law then k_∞ is approximately independent of temperature.

Hence from equations (9.33) and (9.36)

$$\varrho = \frac{k_\infty - 1}{k_\infty} - \frac{B^2}{k_\infty}\left[L_1^2 \left(\frac{T_2}{T_1}\right)^{1/2} + L_s^2\right] \tag{9.37}$$

The temperature coefficient at constant density α_d is obtained by differentiating equation (9.37) with respect to temperature, T_2, and so

$$\alpha_d = -\frac{B^2 L_1^2}{2k_\infty}\cdot\frac{1}{T} \tag{9.38}$$

Hence the temperature coefficient due to these effects will always be negative provided that the absorption cross-sections of all materials in the reactor obey the $1/V$ law. If the thermal neutron energies are near to a resonance level for a particular reactor material then an increase in temperature will raise the thermal neutron energy and so bring it into the resonance region. If such a resonance occurs in the neutron absorption cross-sections then this effect will lead to a much larger nega-

tive value for the temperature coefficient. If, however, a resonance occurs in the fission cross-section, then a positive reactivity
change would result due to the increased fission probability.
This could then lead to a positive temperature coefficient.
Before discussing this further, some consideration will be
given to the effects of density changes due to temperature
variations.

As the temperature changes the number of nuclei per unit
volume will alter which in turn effects the macroscopic absorption and scattering cross-sections. In addition changes in
physical size will alter the value of the geometric buckling.

The macroscopic cross-sections are directly proportional to
material densities and therefore the slowing-down area and
diffusion area will be inversely proportional to the square of
the density. Hence as the migration area $M^2 = L^2 + L_s^2$ we get

$$M_2^2 = M_1^2 \left(\frac{d_1}{d_2}\right)^2 \tag{9.39}$$

where d is the density and subscripts 1 and 2 refer to temperatures T_1 and T_2.

Using equation (9.33)

$$\varrho = \frac{k_\infty - 1}{k_\infty} - \frac{B^2 M_1^2}{k_\infty} \left(\frac{d_1}{d_2}\right)^2 \tag{9.40}$$

The temperature coefficient α_B at constant buckling is
obtained by differentiating with respect to temperature and
will be

$$\alpha_B = \frac{2B^2 M_1^2}{k_\infty} \cdot \frac{d_1^2}{d_2^3} \cdot \frac{\partial d_2}{\partial T_2} \tag{9.41}$$

assuming that the microscopic absorption cross-sections (dealt
with previously) are unchanged.

Allowing for the thermal expansion of the materials, then

$$d_2 = \frac{d_1}{[1+a(T_2-T_1)]^3}$$

where a is the thermal coefficient of linear expansion.

Hence $\qquad \dfrac{\partial d_2}{\partial T_2} = -\dfrac{3ad_1}{[1+a(T_2-T_1)]^4}$ (9.42)

To determine α_B at any temperature T, put $T=T_1=T_2$ and $d = d_1 = d_2$ and substitute (9.42) into equation (9.41),

i.e. $\qquad\qquad \alpha_B = -\dfrac{6B^2M^2}{k_\infty}\cdot a$ (9.43)

Hence the effect of a density change at constant buckling will produce negative values for the temperature coefficient.

In order to consider the effect of changes in buckling equation (9.33) is differentiated with respect to B assuming all other quantities remain constant,

i.e. $\qquad\qquad \dfrac{\partial \varrho}{\partial B} = -\dfrac{2BM^2}{k_\infty}$ (9.44)

If we consider the effect of the radial linear expansion of a cylindrical reactor then

$$r = r_0[1+a(T-T_0)]$$

where r is the radius, and

$\qquad a$ the coefficient of expansion, and

$$B^2 = \frac{j_0^2}{r^2}$$

So $\qquad\qquad \dfrac{\partial r}{\partial T} = ar_0$

and $\qquad\qquad \dfrac{\partial B}{\partial T} = -\dfrac{j_0}{r^2}\dfrac{\partial r}{\partial T} = -\dfrac{j_0 a}{r}$ (9.45)

Thus, the temperature coefficient due to changes in the geometric buckling with other quantities constant is given by

$$\alpha_c = \frac{\partial \varrho}{\partial B} \cdot \frac{\partial B}{\partial T} = + \frac{2j_0^2 \, M^2}{k_\infty} \cdot \frac{a}{r^2} \qquad (9.46)$$

Hence in this case a positive temperature coefficient is obtained.

However, in most reactors the actual change in geometrical size of the reactor materials is very small and therefore α_c does not contribute much to the total temperature coefficient.

The value of the total temperature coefficient can vary over a wide range due to the relative importance of cross-sectional changes and density changes represented above by α_d and α_B.

Water-moderated reactors have a very large negative temperature coefficient due to the large density changes taking place as the water temperature increases. The contribution made by α_B is large in this case. However, if water is used only as a coolant then an increase in temperature will produce a decrease in density and a loss of absorber in the reactor. This will cause an increase in reactivity which may be sufficiently large to give a positive value for α_B. Many reactors use water as both coolant and moderator and the combination of the above effects gives rise to a negative temperature coefficient of a lower magnitude than the water-moderated reactors.

The effect of density changes on graphite-moderated reactors is small. The main contribution to the temperature coefficient is due to changes in absorption cross-sections of the reactor materials. One important aspect in the case of a natural uranium–graphite-moderated reactor is the variation of the graphite temperature coefficient with irradiation. The concentration of plutonium in the fuel builds up with irradiation. The fission cross-section of plutonium-239 has a resonance level near to the thermal energy region. An increase in the neutron temperature in the moderator increases the probability of fission and therefore increases the core reactivity. Hence as plutonium-239 builds up in the core, the graphite temperature coefficient becomes more positive. This increase continues throughout the life of the reactor and a value of $+14 \times 10^{-5}$ per °C can be expected after an irradiation of 3000 MWD/T_e. The increase in graphite temperature coefficient with irradiation is indicated in Fig. 9.6.

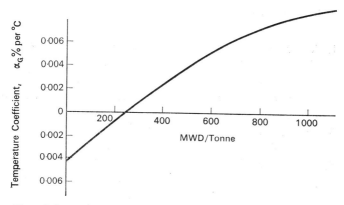

FIG. 9.6. Variation of graphite temperature coefficient with irradiation

9.8.1. The Measurement of Temperature Coefficients at Power

It is important to measure values for the temperature coefficients during operation at power especially in reactors where the coefficient becomes more and more positive with irradiation. Temperature coefficients are required to enable a reactivity assessment to be made and to deduce the magnitude of the reactivity change to be expected with temperature variation. This latter effect may alter the complete mode of operation of the reactor.

It is much easier to measure the overall or combined temperature coefficient of reactivity during operation rather than attempt to measure the individual effects due to the moderator and the fuel. This is quite adequate in most cases. However, for graphite-moderated reactors it is usual to carry out from time to time measurements of the separate coefficients as well as the combined coefficient. The methods to be adopted are as follows.

(a) *Combined temperature coefficient*

The following procedure is adopted.

1. Allow the reactor to stabilise at a particular power and given set of temperature conditions.
2. Carry out a complete temperature survey of the fuel element, moderator and coolant channel outlet temperatures.
3. Measure the neutron flux distribution.
4. Determine the values of $\bar{T}_u$ and $\bar{T}_G$.
5. Reduce or raise the coolant inlet temperature by approximately 20°C. This may be done by altering the rate of L.P. steam dumping or some other suitable means.
6. Maintain the reactor power level as constant as possible by movement of the control rods. The rods to be moved should be chosen so that the minimum alteration of the flux distribution occurs.
7. Carry out a complete temperature survey when conditions have stabilised at the new temperatures.
8. Alter the coolant inlet temperature to re-establish the original temperature conditions and maintain a constant power level during the change.

The reactivity changes due to the variation in reactor temperatures are deduced from the movement of the control rods to maintain a balance. Hence the combined temperature co-efficient may be estimated.

(b) *Separate temperature coefficients*

The following procedure is adopted.

1. Allow the reactor to stabilise at a given set of conditions.
2. Carry out a complete temperature survey.
3. Measure the neutron flux distributions.
4. Reduce the fuel element temperature by rod movement at a rate of approximately 2°C/min to achieve a total temperature change of 100 to 150°C. At the same time maintain the coolant flow and the reactor inlet temperature as constant as possible. The rods to be moved should again be chosen so as to minimise the change in flux distribution.
5. During the temperature change a fuel element temperature survey should be carried out continuously.
6. After the reduction in fuel element temperature the power level is kept constant by rod movement as the graphite moderator temperatures reach equilibrium.
7. The graphite temperature coefficient is deduced from the measured reactivity change. The uranium coefficient is then estimated from a previously measured combined temperature coefficient.

In both these measurements corrections must be applied due to:

(a) The effect of temperature on control rod worths.
(b) The effect due to changes in the neutron flux distribution.
(c) The effect due to any change in xenon concentration.
(d) Any other reactivity change not due to temperature changes in the graphite or the fuel.

9.9. CONTROL ROD CALIBRATIONS AT POWER

If a complex or variable control rod system is installed in order to allow for long-term reactivity changes and flux flattening and shaping during operation, then it is not always possible to calibrate every possible combination and pattern of rods during the nuclear commissioning phase. A different set of control rods may be used and some method of obtaining a calibration at power is desirable. In any case a check on the control rod calibrations should be carried out periodically as so many routine measurements at power depend on these results. A method has been developed which enables a calibration to be carried out at power.[8] This is known as the xenon poisoning technique and is based on the fact that following a reactor shut-down reactivity changes occur due to the variation of the xenon poisoning concentration in the reactor. The technique is to shut down the reactor to a suitable low power level and maintain the power at this level by control rod movement as the xenon poisoning varies. Initially, of course, there is a decrease in reactivity as xenon builds up to its peak value and this is followed by an increase in reactivity as the poisoning decays (see Section 9.5).

The control rod calibration usually commences at the peak of the xenon concentration and consists of inserting the rods to be calibrated to compensate for the release of reactivity as the xenon decays.

The reactor must be operated at power for 2 to 3 days prior to carrying out this technique in order to allow the equilibrium value of the xenon to be reached. The power is reduced to 50 kW for the measurements. The position of the control rods is plotted against time to obtain the calibration. It is then necessary to determine the reactivity change corresponding to the control rod movement. This is done by determining the amount of reactivity released per unit time as the xenon decays. Hence a time coefficient of reactivity is deduced.

The technique employed is to measure the reactor doubling time at different periods during the experiments with the control rods in identical positions. The difference in doubling time will enable the change of reactivity with decay time to be estimated At a suitable stage the control rods are withdrawn to a position which will give a reactor doubling time of approximately 80 sec. The doubling time is measured by the normal methods (see Chapter 6). At the end of the doubling time the control rods will be inserted again to maintain a balance at 50 kW. A second doubling is measured after a period of 1 or 2 hr with the rods withdrawn to the position they occupied for the previous doubling time measurement. The time coefficient of reactivity is then obtained. Doubling time measurements are carried out in this way at frequent intervals during the experiment to cover the whole range of control rod movement. The measured values of the time coefficient of reactivity are plotted against time. Hence using this graph and the rod movement against time graph a control rod calibration may be obtained. The experiment occupies a period of approximately 2 days.

REFERENCES

1. *Transactions of American Nuclear Society* 1960/1966. Sections (a) Reactor operating experience; (b) Reactor operations; (c) Reactor engineering.
2. Calculated surface temperature for nuclear systems and analysis of their uncertainties. *U.S.A.E.C. Research and Development Report* IDO-16343 (1958).
3. GLASSTONE, S. *Principles of Nuclear Reactor Engineering*. Macmillan (1956).
4. STEVENS, C. Confidence limits for the range of a search radar. *Applied Statistics*, Nov. (1957).

5. GRAY, A. L. Measuring flux distributions in power reactors. *Nucl. Power*, April (1958).
6. KLICKMAN, A. E. and DEFALCO, F. R. A wire activation technique for reactor flux profile measurements. *Advances in Nucl. Eng.* Vol. II. Pergamon (1957).
7. KOCH, L., *et al.* New method of measuring neutron fluxes in atomic reactors. Geneva Conf. Paper 15/P/1207 (1958).
8. HOTCHEN, E. P., *et al.* The measurement of reactivity parameters on operational reactors. Conf on Phys. of Graphite Mod. Reactors. Inst. of Physics and Phys. Soc., April 1962.
9. Manual for the Operation of Research Reactors. *I.A.E.A. Technical Reports Series* No. 37 (1965).
10. Yankee reactor operating experience. *Nuclear Safety* **4,** 1 (1962).
11. S.M.1 pressurised water reactor operating experience. *Nuclear Safety* **4,** 2 (1962).
12. Operating experience of the Oak Ridge research reactor. *Nuclear Safety* **5,** 1 (1963).
13. Operating experience at Indian Point nuclear power station. *Nuclear Safety* **6,** 3 (1965).

Index

DATE DUE